光明城 看见我们的未来

BIAD UFo 建筑工作室

BIAD UFo 工作室／群岛工作室 编
同济大学出版社

前言

6　探寻当代设计思潮
　　构建卓越建筑品质／
　　邵韦平

事实

18　人员组织结构 [1.1]
20　工作地点和项目地点 [1.2]
22　项目进程 [1.3]
24　基地建筑红线和基底面积 [1.4]
26　项目高度 [1.5]
28　项目容积率 [1.6]

特写

34　基地 [2.1]
36　工作模型 [2.2]
38　方案册 [2.3]
40　工程记录 [2.4]
42　影像记录 [2.5]
44　游学路线图 [2.6]
46　计划控制表 [2.7]

对谈 UFo

56　徐全胜×UFo [3.1]
60　王辉×UFo [3.2]
64　姜珺×UFo [3.3]
70　尚世睿×UFo [3.4]
76　王舒展×UFo [3.5]

实践

86　数字化：技术自由和思想自由／李淦
90　凤凰中心 [4.11]
106　鄂尔多斯满世广场 [4.12]
114　妫河建筑创意产业园规划
　　　国际竞赛方案 [4.13]
120　妫河建筑创意园接待中心 [4.14]
134　珠海市博物馆和城市规划展览馆 [4.15]
138　中国尊 [4.16]
144　银河SOHO [4.17]
152　序列 [4.18]

158　城市化：城市建造／刘宇光
162　CBD核心区公共空间 [4.21]
172　奥体文化商务园区公共空间 [4.22]
180　北京国际图书城 [4.23]
190　北京世纪华侨城 [4.24]
194　华侨城北京总部 [4.25]
198　北京华侨城社区学校 [4.26]
204　中石油总部 [4.27]
210　朝阳区规划展览馆 [4.28]

218　地域化：别处的建筑／刘宇光
222　奥林匹克公园中心区下沉花园 [4.31]
230　北京图书大厦二期 [4.32]
234　西双版纳机场航站楼 [4.33]
238　巴塘人民小学宿舍 [4.34]
246　中国驻澳大利亚大使馆 [4.35]
252　中国驻印度大使馆 [4.36]
258　BIAD休息亭 [4.37]

同行者

268　OMA建筑事务所 [5.1]
270　盖里事务所 [5.2]
272　扎哈·哈迪德建筑师事务所 [5.3]
274　KPF建筑事务所 [5.4]
276　华汇设计（北京）[5.5]
278　SAKO建筑设计公社 [5.6]
280　KCAP建筑与规划事务所 [5.7]
282　景观都市主义工作室 [5.8]
284　BIAD第四建筑设计院结构设计团队 [5.9]
286　Speirs & Major建筑事务所 [5.10]
288　Christian Kerez建筑事务所 [5.11]

数据库

294　项目信息 [6.1]
305　参展信息 [6.2]
307　摄影师信息 [6.3]
308　团队成员 [6.4]

Foreword

Explore Contemporary
Design Trend & Establish
Excellent Architecture
Quality/ Shao Weiping

Facts

Organization [1.1]
Working Places and
Project Sites [1.2]
Project Progress [1.3]
Red Line Scope and
Foot Print [1.4]
Height [1.5]
Plot Ratio [1.6]

Records

Sites [2.1]
Study Model [2.2]
Design Booklet [2.3]
Construction Images [2.4]
Videos [2.5]
Fieldtrip [2.6]
Schedule [2.7]

Interview with UFo

Xu Quansheng × UFo [3.1]
Wang Hui × UFo [3.2]
Jiang Jun × UFo [3.3]
Shang Shirui × UFo [3.4]
Wang Shuzhan × UFo [3.5]

Practice

86 Digitization: Freedom of Technology and Ideas/ Li Gan

90 Phoenix Center [4.11]
106 Erdos Manshi Square [4.12]
114 Beijing Gui River Architecture Innovation Park [4.13]
120 Show Room of Beijing Gui River Architecture Innovation Park [4.14]
134 Zhuhai City Museum and the Urban Planning Exhibition Hall [4.15]
138 China Zun [4.16]
144 Galaxy SOHO [4.17]
152 Sequence [4.18]

158 Urbanization: Urban Construction/ Liu Yuguang

162 CBD Core Area Public Space [4.21]
172 Olympic South: Culture Zone, Business Park and Public Space [4.22]
180 Beijing International Book Mall [4.23]
190 Beijing OCT [4.24]
194 HQ of OCT Group Beijing [4.25]
198 School of Beijing OCT [4.26]
204 Headqarter of China National Petroleum Corporation [4.27]
210 Chaoyang District Urban Planning Exhibition Hall [4.28]

218 Localization: Architecture in Other Places/ Liu Yuguang

222 Sunken Garden in the Center Area of Olymipc Green [4.31]
230 The Second Phase of Beijing Book Building Project [4.32]
234 Xishuangbanna Airport New Terminal [4.33]
238 Dormitory Design of Batang School Campus [4.34]
246 Chinese embassy in Australia [4.35]
252 Chinese embassy in India [4.36]
258 BIAD Pavilion [4.37]

Cooperators

268 Office for Metropolitan Architecture [5.1]
270 Gehry Partners LLP [5.2]
272 Zaha Hadid Architects [5.3]
274 Kohn Pedersen Fox Associates [5.4]
276 HHD_FUN [5.5]
278 SAKO Architects [5.6]
280 KCAP Architects & Planners [5.7]
282 LAUR Studio [5.8]
284 Structure Design Team, BIAD Architectural Design Division No.4 [5.9]
286 Speirs and Major Associates [5.10]
288 Christian Kerez Zurich AG [5.11]

Database

294 Projects [6.1]
305 Exhibitions [6.2]
307 Photo Credits [6.3]
308 UFo Members [6.4]

前言
探寻当代设计思潮 构建卓越建筑品质/邵韦平

我们是谁

BIAD UFo工作室是中国最具影响力的民用设计机构——北京市建筑设计研究院有限公司中的核心设计团队。作为UFo工作室领导的本人同时还担任着BIAD集团的执行总建筑师职务,因此工作室同时肩负着自身发展和提升BIAD品牌水准的双重目标。近年来我们承担了许多极具挑战性的设计项目,完成了一批高质量的设计作品;团队经历十年磨炼,大浪淘沙,逐步培养出一批具有优秀专业素养、敏锐设计眼光和高度敬业精神的设计骨干,同时,伴随着品牌提升,吸纳了一批有良好教育背景和积极向上的年轻设计人才。工作室的力量在不断壮大,承接任务的能力也有显著提高。

我们在做什么

BIAD UFo工作室的设计项目涉及城市规划、城市设计、市政设计、建筑设计、室内设计、景观和公共艺术品设计等。长期的设计实践积累和研究感悟,让我们对城市、建筑、文化以及人的需求有了更专业的理解。我们力图克服在传统体制下形成的种种不完整的设计观,尝试建立起一套符合建筑发展规律的设计方法。通过自身的努力,我们独立完成了一批极具挑战性的建筑项目,积累了宝贵的高端设计经验,我们有信心完成各种高难度建筑的挑战。

在过去的十年里,我们与许多世界顶级的建筑大师及团队开展合作,包括弗兰克·盖里、诺曼·福斯特、扎哈·哈迪德、雷姆·库哈斯、KPF、Arup等。与大师同行也让我们有机会深入体会到前沿的设计思想,同时对我们形成更符合当代建筑设计规律的科学设计观与方法产生重要的影响。

Preface
Explore Contemporary Design Trend &
Establish Excellent Architecture Quality / Shao Weiping

Who we are
BIAD UFo is the core design team of the Beijing Institute of Architectural Design Co.,Ltd.(BIAD), the most influential civil design institution in China. As the leader of UFo, I also act as the executive chief architect of BIAD and undertake the responsibilities to achieve the dual goals of self-development and uplifting the brand level of BIAD. Recently, we've undertaken many challenging design projects and completed a number of high-quality design works. Through a decade of tempering and screening, UFo has gradually cultivated a group of key designers with good professional quality, keen design insight and highly professional devotion. With brand improvement, a group of young designers with a good educational background and positive attitude have been attracted into UFo. The strength of UFo is continuously developing and the capacity to undertake tasks has been substantially improved.

What we do
The design projects of BIAD UFo include urban planning, urban design, municipal design, architectural design, interior design, landscape design, design of public art, etc. Based on accumulation of our long-term design practices and research inspiration, we have achieved a more professional understanding of city, architecture, culture, and human needs. Moreover, in an attempt to overcome various incomplete design concepts formed under the traditional system, we have established a set of design methods suitable for the rules of architectural development. Through our efforts, we have independently completed challenging architectural projects and accumulated valuable experience in high-end design. We are fully confident in completing various challenges of highly demanding architecture.
Due to business relationships, we have cooperated with many top international architects in the last decade, including Frank Owen Gehry, Norman Robert Foster, Zaha Hadid, Rem Koolhaas, KPF, Arup, et al. By working with them, we are able to further understand cutting-edge design ideas. Meanwhile, it plays a vital role for our team in forming a rational methodology and design methods suitable for contemporary architectural design rules.

我们主张的设计观

城市与场所精神

城市承载着建筑和人的生活,因此城市问题是建筑设计必须首先思考和应对的挑战,建筑师要出色完成自己的职业任务,就必须了解建筑所在城市的历史,尊重城市发展的规律,用科学精神和当代审美来塑造城市的未来,让新的建筑成为调节城市环境和弥补城市缺陷的积极因素,而不是城市的负担。

建筑设计不是一种可以完全个性化的职业,建筑师必须有社会责任。建筑一旦形成,必然对所在环境产生不可回避的影响。对于一座建筑,最低要求是融入环境,成为建成环境中的和谐因素;更高的要求是为所在人文、自然环境作出有益的贡献,从而提升环境的整体品质。一个好的创意必须基于对所在环境的研究与发掘,这样才能创作出一个真正属于那个场所的建筑,才能找到体现地域文化和场所个性的合理方案。

设计创新与技术美学

现代主义突破了古典建筑的繁复,开创了简约的建筑时代,让建造获得了空前的自由,从而造福于更广大的民众。但随着社会发展和技术的进步,高品质的建筑不能仅仅停留在形式表达上,建筑师不仅要关注建筑形式的创造,更要学会运用建筑技术的语言——材料、技术构件和所有的空间语素——来塑造建筑的整体美,包括形式美与技术美。虽然技术不是建筑学的全部,设计师需要在一个更广泛的框架内对技术进行划界和限制,使之能够丰富和拓展参与者体验的广度和深度,技术构建是通往建筑真实世界的唯一道路,也是面向未来的全部建筑意义所在。

Our Design Philosophy

Urban and Place Spirit
Cities bear architecture and human life, thus, urban issues are a top concern and primary challenge for the architectural design. Architects must understand the history of the relevant city if they are to properly complete their own professional tasks. This includes the area in which the architecture will be located, respecting rules of urban development, and shaping the city's future with scientific spirit and contemporary aesthetics so as to make new architecture form a positive factor for adjusting urban environments and making up for urban defects, rather than place a burden on the city. Architects are obliged to undertake social responsibilities, as architectural design is not a completely individualized profession. Architecture will have an inevitable effect on local environments once they are constructed. The minimum requirement for architecture is to integrate so as to become a harmonious factor in the built environment. Higher requirement is to contribute to human and natural environments in order to improve overall environmental quality. A good idea is always based on research and exploration of the local environment. Only in this way will an architecture really become part of a location and a reasonable proposal representing regional culture and place be hammered out.

Design Innovation and Technical Aesthetics
Modernism breaks through the pedantry of classic architecture and creates a concise architectural age, offering construction unprecedented freedom and bringing public benefits. With social development and technical progress, however, high-quality architecture should not merely stay at formal representation. Architects should not only pay attention to the creation of architectural form, but also learn to utilize the language of architectural technology – materials, technical components and all space morphs to shape the overall beauty of architecture, consisting of beauty in form and beauty of technology. Although technology cannot represent the whole of architecture and designers need to restrict and limit the technology in a more extensive frame to enrich and expand the extent and depth experienced by participants, technology construction is the only access to the real world of architecture and also the meaning of all future-oriented architecture.
Technology has clear marks for methods and interacts through connection with functions and places, realizing coordination with nature and history under a new order. The form quality and meaning are from the process and

技术带有清晰的方法的印记，通过与功能、场所发生关联而相互作用，在全新的秩序下实现与自然、历史的协调。形式的品质和意义来自技术在建筑中的过程和方法，技术需要被谨慎、准确和最大限度地加以利用，以适度的方式在建筑中实现具体的真实性，准确地反映今日世界的复杂状况，在丰富多样的当代文化中获得深刻的意义。

逻辑建构中，通过文化选择和创造性的幻想对抗机械理性，为建筑注入生命的活力和光彩，接近建筑的本质，实现建筑精神的升华和超越，最终在建筑中建立技术与人的自由关系，使技术在建筑学的范畴中融入当代文明的进程。

人性化与精细化设计

建筑是一门关于人类及其生活质量的艺术。建筑与城市一样，要满足最终使用者的生理与心理、物质与精神、个人与社会、当今与未来等对建成环境的需求。设计不只是关注建筑的物质性特征，还要从更高层次关注人的心理体验，关注建筑与人的身体行为效果之间的关系。

在科技发达的今天，人性化的另一层含义是精致性，即通过更加细致的设计和精确的建造，让建筑更周到地服务于人，满足人不同层次的需求。在使用者所有可达到、可触及、可观察的范围内，创造出周到、精确的建筑细节，来满足当代人对现代生活品质和审美的需要。

文化传承与建筑当代性

建筑的价值不仅体现在实用功能方面，还在于文化传承作用，而且文化可以被建筑长久地体现，影响着人类的生活与发展。为了塑造地域个性，建筑师应该从文化传统中汲取营养，增加建筑的文化附加值。但传承文化并不意味着机械复制传统符号——优秀的建筑应该是从传统中提炼符合当代价值的内容，可同时经受现代审美和传统精神的双重考验，并能够影射出建筑未来的远景。

method of technology in architecture. Technology needs to be utilized to the most extent and in a prudent and accurate fashion to realize actual reality in architecture in a proper manner, and to correctly reflect the complicated situation of the current world and obtain profound meaning from a rich variety of modern cultures. In logical construction, we will utilize cultural selection and creative fantasy to oppose mechanical rationality to inject vitality into architecture, so as to approach the essence of architecture and realize sublimation and transcendence of the architectural spirit. Ultimately, a free relationship between technology and humanity in architecture would be established to integrate technology into the process of modern civilization in the category of architecture.

About Humanity and Building Quality
Architecture is the art relating to humanity and its life quality. As with a city, architecture shall meet physiological and psychological, material and mental, individual and social, present and future needs of end users on the built environment. Architectural design does not only focus on material features, but also on psychological experience of humans at a higher level and relationships between architecture and the effects of people's physical behaviors.
Nowadays, with advanced science and technology, the implication of humanization is exquisiteness in other aspects. It means that architecture shall serve people more attentively to meet the needs of different levels through more exquisite design and accurate construction. For accessibility to users, considerate and accurate architectural details shall be created to meet the aesthetic needs of humans on modern life quality.

Cultural Heritage and Architecture Contemporariness
The value of architecture is not only reflected in the utility function, but also in the function of cultural heritage. Meanwhile, culture can be reflected by architecture through a long-term influence on human life and development. To shape regional features, architects should absorb nutrition in cultural tradition to increase the added value of architectural culture. Cultural heritage doesn't mean copying traditional signs in a mechanical manner. Remarkable architecture should contain content suitable for the modern value extracted from tradition and withstand the dual challenges of modern aesthetics and traditional spirit, as well as reflect future prospects of the architecture.

可持续发展策略

建筑所消耗的自然资源占人类消费自然资源的比例是一个十分惊人的数字。保护自然环境、减少资源消耗、保持生态平衡的可持续发展思想已经成为当今建筑界的共识。建筑设计不仅要考虑市场需求和个性张扬,还要考虑公共利益和可持续发展的可能。

可持续发展思想下的绿色设计作为一门通用建筑技术正逐步走向成熟,作为未来设计实践不可或缺的内容得到广泛应用。绿色设计既是技术,也应该是一种设计哲学,应该成为建筑师必不可少的修养,以便引领其职业活动。设计既要考虑眼前的需求,也要关注未来长远的发展可能;既要有具体的绿色技术,也要有整体系统的可持续发展策略。绿色设计不是孤立的,它需要成为一种信念,融入专业活动的全过程。

我们的职业目标

建筑是一个有悠久历史的行业,建筑是人类文化的重要载体,承载着人类数千年灿烂的历史文明。但由于建筑受到多种客观因素制约,相对于高端的现代制造业,建筑业一直处于较为粗放的状态,这种情况在相对欠发达的中国城市更为突出。20世纪后期以来的几十年,随着人们对自然世界认知能力的提高,随着建筑科技的进步和物质水平的改善,当代建筑呈现出空前的繁荣。BIAD UFo工作室正是在社会与科技高速发展的背景下走过了十年的发展路程。在过去的十年里,我们始终保持开放的专业心态和国际化的视野,努力摒弃过时的设计观念,探索当代设计的发展趋势,用现代设计方法和执着的专业热情,成就了一批对城市和环境有积极影响的建筑作品,同时也形成了团队的风格与个性。我们倡导从场所环境和人的需求中汲取灵感,运用整体设计思想和现代建构方法,让技术营造出建筑的精神价值,从而为业主提供符合当代审美标准的高品质的建筑文化产品。

Sustainable Development Strategy

It is striking the proportion of natural resources consumed by architecture accounting for those by humans. The idea of sustainable development consisting of the protection of natural environment, reduction of resource consumption and ecological balance has already become a consensus in the modern building industry. For architectural design, it is not only market demands and publicizing individuality that need to be taken into account, but also public interest and the possibility of sustainable development to be considered.

As a common architectural technology, green design under the idea of sustainable development is gradually moving toward maturity, and has been applied extensively as an indispensable content for future design practice. Green design is technology and also shall be a kind of design philosophy, which shall become the indispensable culture for architects to guide professional activities. For design, present demand and the possibility of future development should both be focused on. There should be specific green technology and also the strategy of sustainable development as a whole system. Green design is not isolated, it has to be a faith integrated into the whole process of our professional activities.

Our Vision

Architecture is an industry with a long history and an important carrier for human culture, bearing the bright historical civilization of humanity for thousands of years. Limited by various objective factors, however, architecture has always been in an understated condition compared with the high-end modern manufacturing industry, which is more obvious in relatively underdeveloped cities in China. Since the end of the twentieth century, with the improvement of human cognitive competence of the natural world and progress in architectural technology and enhancement of material level, modern architecture is booming at an unprecedented rate. BIAD UFo has gone through a decade of development under the background of high-speed development of society, science and technology. During that period, we have persistently kept an open professional attitude and an international vision, made efforts to abandon backward and outdated design ideas, explore development trends of modern design, and create a number of buildings with positive effects on city and environment by modern design methods and persistent professional enthusiasm, as well as forming the style and individuality of our team. We advocate the absorption of inspiration from the environment, human needs and utilization of the whole design idea and modern construction methods to empower technology with the spiritual value of architecture, thus providing owners with high-quality architectural culture products compliant with modern aesthetic standards.

事实

Facts

人员组织结构 [1.1]
工作地点和项目地点 [1.2]
项目进程 [1.3]
基地建筑红线和基底面积 [1.4]
项目高度 [1.5]
项目容积率 [1.6]

Organization 1.1
Working Places and Project Sites 1.2
Project Progress 1.3
Red Line Scope and Foot Print 1.4
Height 1.5
Plot Ratio 1.6

人员组织结构

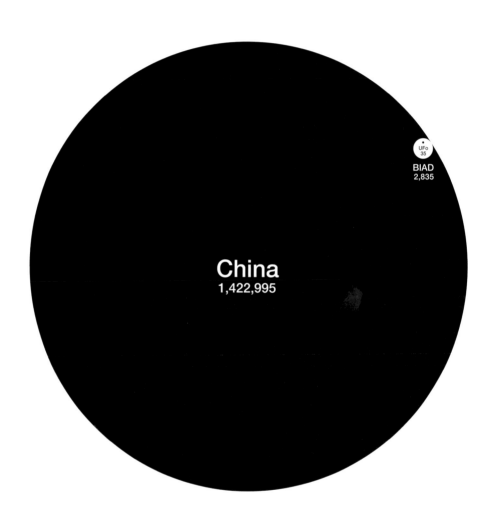

Organization

1.1

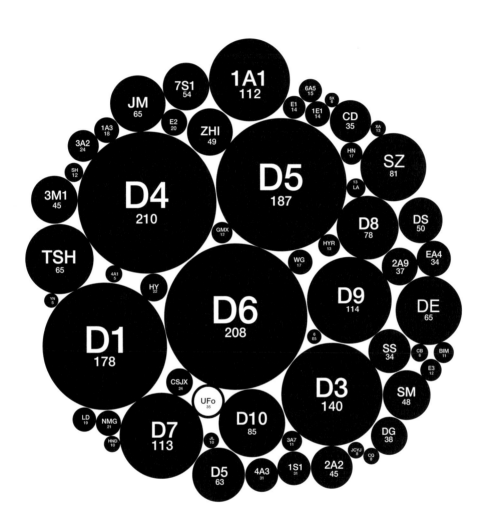

(左页)中国勘察设计机构从业人员与BIAD及UFo的员工数量比较;(右页)BIAD设计团队规模比较
(left)Comparison of employees of China's exploration & design institutes with BIAD and UFo. (right)Scale comparison of BIAD's design teams

工作地点和项目地点

Working Places and Project Sites

1.2

UFo团队曾经工作地点和项目地点在全球的分布
Global distribution of UFo's previous working sites and project sites

项目进程

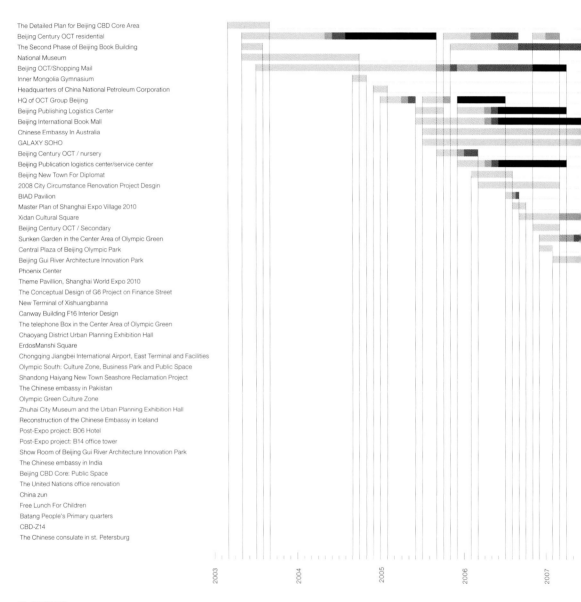

Project Process

1.3

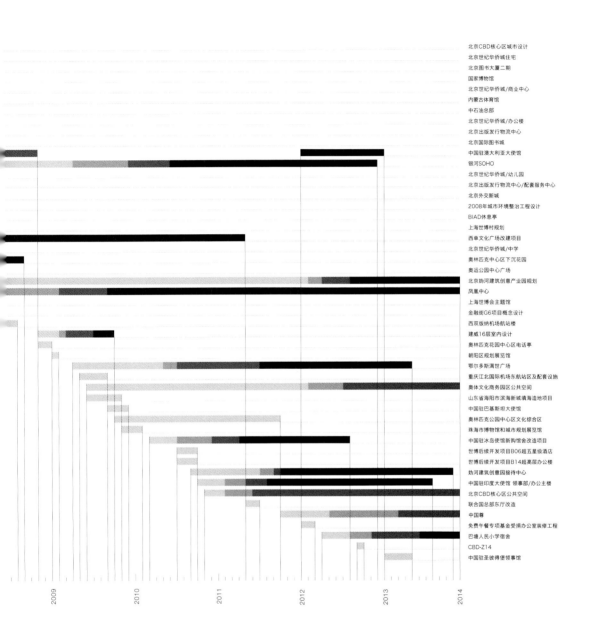

UFo历年项目的各阶段进展
Progress of UFo projects at each stage over the years

基地建筑红线和基底面积

奥体文化商务园区公共空间
Olympic South: Culture Zone, Business Park and Public Space

93,102
281,869

妫河建筑创意产业园规划国际竞赛方案
Beijing Gui River Architecture Innovation Park

53,665
210,150

巴塘人民小学宿舍
Dormitory Design of Batang School Campus

1,915
30,560

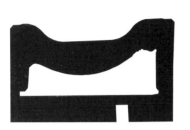
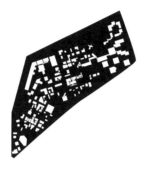

北京图书大厦二期
The Second Phase of Beijing Book Building

5,010
11,698

鄂尔多斯满世广场
Erdos Manshi Square

7,830
22,804

朝阳区规划展览馆
Chaoyang District Urban Planning Exhibition Hall

9,967
13,380

中国尊
China Zun

5,000
11,478

北京华侨城社区学校
School of Beijing OCT

3,115
20,900

华侨城集团北京总部
HQ of OCT Group Beijing

1,580
3,004

Red Line Scope and Foot Print / m²

1.4

CBD 核心区公共空间
CBD Core Area Public Space

103,670
113,172

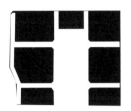

北京国际图书城
Beijing International Book Mall

19,500
241,330

西双版纳机场航站楼
Xishuangbanna Airport New Terminal

15500
102,500

中石油总部
Headqarters of China national petroleum corporation

9,321
22,519

凤凰中心
Phoenix Center

7,278
18,821

奥林匹克中心区下沉花园
Sunken Garden in the Center Area of Olympic Green

6,394
45,000

CBD-Z14
CBD-Z14

1,546
2,427

妫河建筑创意产业园接待中心
Show Room of Beijing Gui River Architecture Innovation Park

1,361
2,957

UFo 部分项目的基地建筑红线范围和建筑基底面积比较
Comparison of red line scope and foot print for UFo projects

项目高度

20.6
8- 北京华侨城社区
8- School of Beijing

13.7
7- 巴塘人民小学宿舍
7- Dormitory Design of Batang School

10.8
6- 朝阳区规划展览馆
6- Chaoyang District Urban Planning Exhibition Hall

9.8
5- 妫河建筑创意产业园接待中心
5- Show Room of Beijing Gui River Architecture Innovation Park

9.3
4- 华侨城集团北京总部
4- HQ of OCT Group Beijing

-6.15
3- 奥林匹克中心区下沉花园
3- Sunken Garden in the Center Area of Olympic Green

-18
2- 奥体文化商务园区公共空间
2- Olympic South: Culture Zone, Business Park and Public Space

-23.8
1- CBD 核心区公共空间
1- CBD Core Area Public Space

1.5

528
16- 中国尊
16- China Zun

111.5
15- 鄂尔多斯满世广场
15- Erdos Manshi Square

90
14- 中石油总部
14- Headquarters of China national petroleum corporation

69
13- CBD-Z14
13- CBD-Z14

52
12- 凤凰中心
12- Phoenix Center

38
11- 北京图书大厦二期
11- The Second Phase of Beijing Book Building

34
10- 西双版纳机场航站楼
10- Xishuangbanna Airport New Terminal

Book Mall

项目容积率

中国尊
China Zun

30.5

CBD-Z14
CBD-Z14

6.5

中石油总部
Headquarters of China national petroleum corporation

6.36

北京国际图书城
Beijing International Book Mall

3.4

CBD 核心区公共空间
CBD Core Area Public Space

3.3

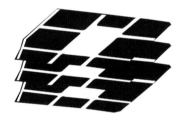

朝阳区规划展览馆
Chaoyang District Urban Planning Exhibition Hall

0.97

妫河建筑创意产业园规划国际竞赛方案
Beijing Gui River Architecture Innovation Park

0.87

妫河建筑创意产业园接待中心
Show Room of Beijing Gui River Architecture Innovatio

0.81

Plot Ratio

1.6

鄂尔多斯满世广场
Erdos Manshi Square

4.27

凤凰中心
Phoenix Center

2.03

北京图书大厦二期
The Second Phase of Beijing Book Building

3.6

华侨城集团北京总部
HQ of OCT Group Beijing

1.4

奥体文化商务园区公共空间
Olympic South: Culture Zone, Business Park and Public Space

1.06

巴塘人民小学宿舍
Dormitory Design of Batang School Campus

1.02

北京华侨城社区学校
School of Beijing OCT

0.57

西双版纳机场航站楼
Xishuangbanna Airport New Terminal

0.32

UFo部分项目的容积率比较
Plot ratio comparison between UFo projects

特写

Records

基地 2.1
工作模型 2.2
方案册 2.3
工程记录 2.4
影像记录 2.5
游学路线图 2.6
计划控制表 2.7

Sites [2.1]
Study Model [2.2]
Design Booklet [2.3]
Construction Images [2.4]
Videos [2.5]
Fieldtrip [2.6]
Schedule [2.7]

基地

Sites

2.1

UFo 部分项目的基地照片和建筑范围
Onsite photo of UFo Projects

工作模型

方案册

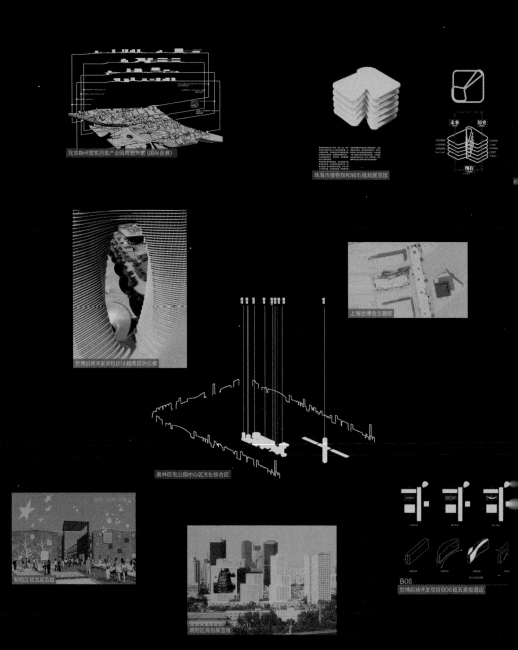

Design Booklet

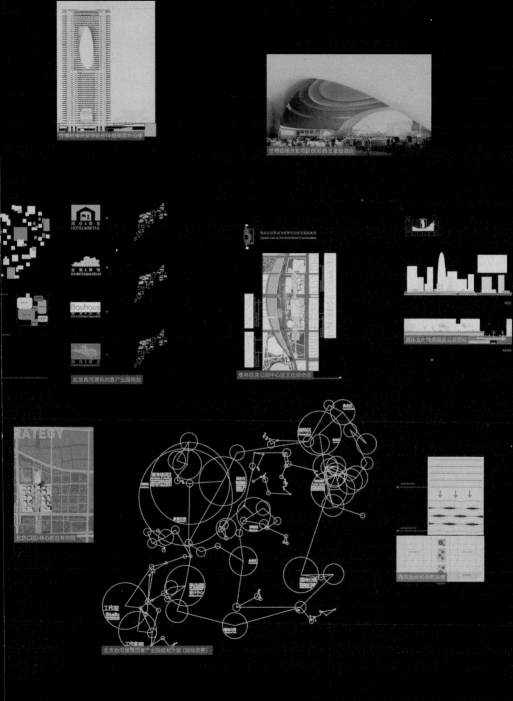

UFo 部分方案图册节选
Pages from UFo proposal books

工程记录

Construction Images

影像记录

Videos

威尼斯双年展展品安装现场

Steven Holl先生在凤凰中心工地

在丁垚老师导引下参观独乐寺

和FOSTER公司设计师讨论国家博物馆竞赛方案

UFo团队考察俄罗斯圣彼得堡项目访问当地设计院

团队及哈佛RAFI教授讨论CBD核心区城市设计

上海世博会主题馆方案推敲

SOHO/ZHA/BIAD银河SOHO项目WORKSHOP

年会现场

帅大厦二期工程

2007深圳香港双城双年展论坛

在驻澳大利亚使馆讨论设计方案

游学路线图

Fieldtrip

2.6

UFo 历次学习考察的地点和人次
Location and participant number of previous Fieldtrip

计划控制表

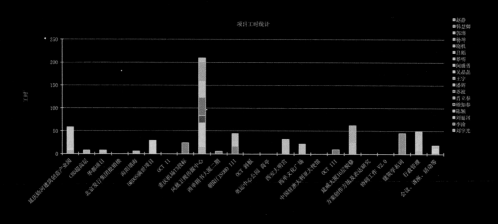

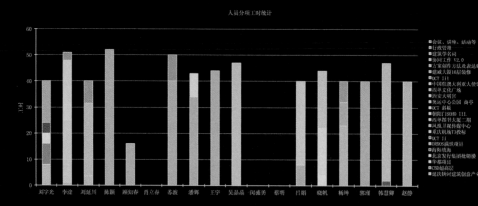

Schedule

2.7

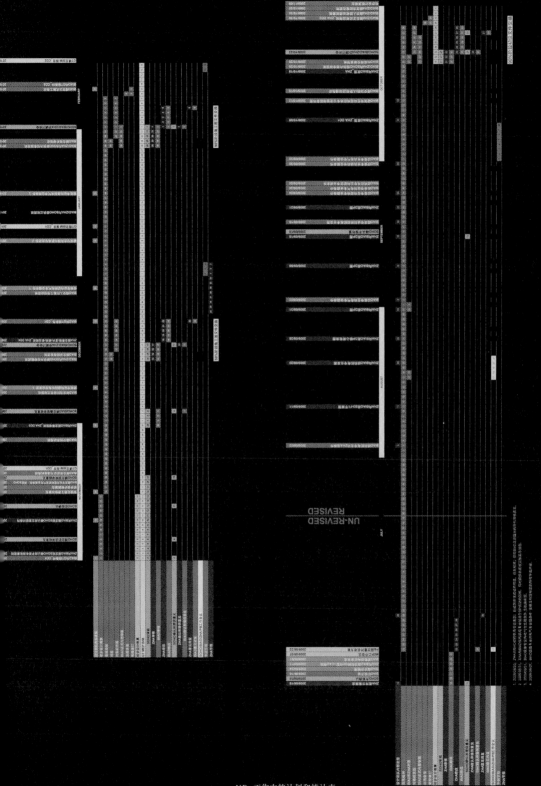

UFo 工作中的计划和统计表
Schedule and statistical tables during working procedures

对谈UFo

Interview with UFo

徐全胜 ×UFo [3.1]
王辉 ×UFo [3.2]
姜珺 ×UFo [3.3]
尚世睿 ×UFo [3.4]
王舒展 ×UFo [3.5]

Xu Quansheng × UFo 3.1
Wang Hui × UFo 3.2
Jiang Jun × UFo 3.3
Shang Shirui × UFo 3.4
Wang Shuzhan × UFo 3.5

对谈 UFo

Interview with UFo

UFo的发展一直伴随着与各领域人士的交流与合作,这些交流合作拓展了UFo的视野和感知力,为UFo的发展注入了多元化的力量。在这一章节中,我们邀请了来自国营设计院、独立事务所、评论界、业主方和专业媒体领域的五位代表人物——徐全胜、王辉、姜珺、尚世睿和王舒展,分别与UFo负责人邵韦平、刘宇光及李淦展开对谈,就UFo以及中国设计机构的现状、发展等话题进行了广泛的交流。

Collaboration with professionals from various fields has played a significant role in the development of UFo, expanding our horizon and infusing diversity. In this chapter, we invite Xu Quansheng, Wang Hui, Jiang Jun, Shang Shirui and Wang Shuzhan, representing respectively state-owned design institutes, independent offices, critics,client and professional media, to talk with Shao Weiping, Liu Yuguang and Li Gan - three partners of UFo, on topics regarding the status quo and development of UFo and design institutions in China.

徐全胜 × UFo

徐全胜，北京市建筑设计研究院有限公司董事、总经理、副总建筑师、国家一级注册建筑师、教授级高级建筑师、中国建筑学会建筑师分会理事、清华大学建筑学硕士

Xu Quansheng, Director, General Manager and Deputy Chief Architect, BIAD. National First-Class Registered Architect, Professor-Level Senior Architect. Member of Architect's Branch, Architectural Society of China. M.Arch of Tsinghua University

3.1

邵韦平　　　　UFo与北京院的其他综合所不太一样，是一个单专业、相对自主运行的团队，没有过多的包袱和陈规戒律，可以全身心地投入到设计中来，不被琐事干扰，这也是我们今天能有较好发展的前提条件。今天请徐总过来，一起探讨一下北京院这些年给我们建筑师提供了怎么样的专业发展条件。2002年朱小地院长跟我谈，计划组织一个班底，做一些超越普通商业项目，具有研究性和前瞻性的工作。最早做中央台的时候，我们还没有形成机构呢，只是在一幢住宅楼里占了两套房子，大家凑在一起先干活。后来变成一个工作室，经过十年的发展和变化，大家沉淀下来了，所有留下来的人都成为特别优秀的员工。总的来说还是利用了当年这个机会，有了这些发展。

徐全胜　　　　这些年社会高速发展，建筑设计这个行业也有很多机遇。你们是北京院平台上的一个部门，叫"方案创作工作室"。"创作"这两个字强调了它的属性。当然，叫"创作"并不是单指概念设计，全过程都可以是创作。

邵韦平　　　　我们现在希望把"方案"两个字去掉，改叫"创作研究工作室"，不然这名字有歧义，好像创作仅停留在方案阶段，其实我们的工作是贯穿工程的全过程。

徐全胜　　　　这些可以通过作品来证明。刚才在说大的机遇，我认为也有小的机缘。邵总和李淦、刘宇光，你们三人形成了明星设计师核心层，有鲜明、一致的风格和价值观，BIAD拥有丰富的人力资源，以你们为领军人物，在BIAD的平台上聚集有共同追求的设计师，从而形成了UFo团队。

邵韦平　　　　这是很多院外团队非常羡慕的一点，齐欣大师就说，小公司干不了如此复杂的活。

徐全胜　　　　对于UFo的设计项目，配合的团队都积极主动地对待。每个工程都有创作和设计上的追求以及技术上的难度，大家配合得不错，当然也会有一些问题。不光是建筑，各专业都上了一个台阶。这也是院里希望看到的，在以市场和生产为导向的公司里有一个部门专注做一些具有行业引领性的工作。那个时候我是四所所长，四所是UFo的的主要合作者之一。

邵韦平　　　　凤凰那个项目如果不是徐总的鼎力支持，好多事情不会这么顺利。四所一直是先干活后算账。

徐全胜　　　　品牌的价值是无限的，对我们来说，凤凰卫视、银河SOHO这些项目的设计费虽然不高，但建成之后的社会效果和带来的潜在价值是无法计量的，会远远超过设计费的价值。虽然邵总把机遇当成第一个重要话题提出来，但我觉得机遇对于UFo取得的成就而言，只是因素之一，其他团队也都有类似的机遇。大家总认为建筑师在管理上是很自我的,建筑师的这种自我体现在想要三个权利:一是这个事做还是不做;二是设计怎么做;三是找谁来合作。纯粹生产型的部门是没有这个权利的,但设计师渴望这些权利,有了这些权利就很满足。不过做到最后,规模壮大了,就得为下面的人负责,工资、奖金、五险一金、房租水电、模型费等,这些压力大了,(负责人)就必须向商业经理人的角色转换。

邵韦平　　　　　　　　也要考虑生存。要是没有院里的机制,很多事是实现不了的。我们的项目聘请了很多顾问,比如凤凰项目聘请的BIM顾问,所需费用占总设计费的比重是相当高的。如果我不是拍板的人,另外有一个行政所长,他肯定会对这个费用提出质疑。现在回过头来看,钱是花得有价值的,如果没有BIM系统,做出来的可能就不是这个东西,也许是省了钱了,但效果也没有了。另外,如果院里是按工程计奖,设计师会就想这么点收入,怎么能陪你六年呢?我们能这么干活,也是因为北京院高层领导提供了一些条件。

徐全胜　　UFo用十年时间形成品牌,这个速度还是挺快的。北京院用了64年,二、三、四、六所等一些大所也用了几十年的积累。

邵韦平　　　　　　　　UFo经过十年的发展,眼下需要进一步进行自我的定位调整。我们来讨论一下,在北京院乃至整个行业里,如何定位,如何发挥我们的能力。在过去十年里,作为一名体制内的技术带头人,我一直有压力。本土的设计公司和境外的设计公司相比,一直很被动,央视、鸟巢、大剧院这些项目都是外方设计的。地标性建筑一直是外方占优势,中方只是参与一些外方不适合的项目,比如政府项目、办公楼和保密项目。另外,社会上还有一种观点,认为体制内的设计公司缺乏创造力,体制外的建筑师更有活力,有开放视野。这双重压力一直伴随着我们多年。我们的前辈,像张镈那一代人,地位是很高的,在市领导甚至总理那里都是座上宾。而我们这一代人没有这个条件,可能也没有达到他们的境界。我们一直有突破的愿望,建立起信心,使团队乃至中国的设计师有所作为,能够和老外平等地交流。我认为到讨论这个话题的时候了,咱们探讨一下UFo团队在当前环境条件下能不能发挥改变现状的作用。

徐全胜　　这是社会问题。我觉得社会上已经有这方面的声音,最近一些大众媒体已经在关注:为什么中国的大型建筑都是洋面孔在做设计?过去只是建筑学术圈内部讨论的问题,现在已扩大到政府和大众层面了。由此可以判断政治成熟期到了,相应的,机会也会出现。至于我们有没有承担的能力,我觉得是有的,而且在很多方面是超过外国建筑师的。但是,文化、社会背景的差异性,导致国外设计师的设计从方案、过程到完成之后的效果与中国建筑师相比还是存在差异的,这跟好莱坞的电影和冯小刚的电影有区别是一个道理。文化没有高低论,但文化产品的差异性还是有的。目前中国建筑师的能力在逐渐加强,社会地位在不断提高,发挥的作用也在逐渐加大。

邵韦平　　　　　　　　这也是水到渠成的事,摆脱困境也不是仅靠外力就能完成的,主要还是要靠自身能力的提升。

徐全胜　　怎么证明自己的实力呢？还是要靠建成的作品，靠累积的原创或者合作的项目。创新有两个面向，一是像凤凰这样的项目，在国际上也是最前端的，特别是灯光、夜景照明体系，都体现了与国际平行的创新能力。另外，在中国，由于整个社会的不配套，还原设计过程中应有的理论和体系也意味着创新。在香港，外表皮、建筑设计、结构、机电、造价控制可以分给五家公司来做，再加上相关的顾问，涉及面非常广，但相互之间的协作却很顺畅，同时也把决策的危险扁平化了。而我们由于各行业发展的不均衡，专业方、配套方、总包方、供货商、专项技术都参差不齐，导致建筑师得自己想办法去发现好的团队，继而协作。

朱院长对于UFo的定位，是具有超大型建筑设计示范作用的北京院品牌工作室；邵总坚持要成立建筑基础理论研究室，是对两种面向创新的回应。在改革开放前，社会及市场环境稳定，几位总建筑师形成了类似的价值观，对于建筑的功能、比例、尺度等基本问题总结了一套方法，通过言传身教，形成了BIAD的态度、观点、设计理论和方法。目前，BIAD从设计师到设计团队，价值观是多元的，你们现在的工作就是给北京院梳理出一套大型建筑设计的教材和价值观，接下来我们的工作是要把它传递到北京院其他设计部门去。这个工作很紧迫。朱院长总是说，要搞明白自己在建筑行业中的身份，要在言谈举止、品质品位、观点态度上都体现出北京院的特点来。你们的工作就是在为这些特点下定义。

邵韦平　　希望通过我们的工作，为北京院在社会上塑造一种正面、专业的形象。社会上确实有一些人对体制内设计院有成见，认为我们是官办的，靠垄断生存；我们要用我们的答卷扭转这种成见。刚才说机遇，还漏了一点。过去十年我们有很多与国外顶级公司合作的机会，如福斯特、OMA、扎哈、盖里等，这也是我们团队进步比较快的原因之一。这些机会比在国外读个学位还要直接，工作室每个人都获得了建筑养分，得到了专业提升。还有就是新技术的机遇，让我们克服了地域的屏障，国内外同行汇集在同一个环境里。当然，最后提升多少还在于个人的努力。

徐全胜　　中国建筑师发展的瓶颈不在能力，在于素质。一是对于技术的严谨务实，以及学习的方法能力；二是文化艺术修养。

邵韦平　　最后请徐总给我们展望一下未来。

徐全胜　　十年了，UFo从团队到业绩到品牌都有了积累，下一步要有选择地做一些事，毕竟精力是有限的。在原创上需要进一步努力，在这方面更有些锋芒和力度。另外，要把成果物化，让其他团队也能受益，把工作方法、项目管理方法、态度、价值观这些核心竞争力推广到全院，甚至能对中国当代的建筑设计形成好的影响。

邵韦平　　总结一下徐总的希望：有选择性地做更多的好建筑，形成一些物化的理论和方法，让更多人共享，培养更多的合格建筑师，创造好的建筑创作氛围，担负起时代赋予我们的职责。

徐全胜　　要超越设计，这是很有意义的。

王辉 × UFo

王辉，URBANUS都市实践创建合伙人、主持建筑师，美国纽约州注册建筑师，
清华大学建筑学士、硕士，美国迈阿密大学建筑硕士
Wang Hui, Founding partner and principal of URBANUS. Registered architect in State of New York ; B.Arch and M.Arch of Tsinghua University; M.Arch of Miami University

3.2

邵韦平　　　　　　　　请您站在局外人的角度对我们工作室提一些看法和意见。

王辉　　那我就站在"if I were you"的角度来评价一下（笑）。在之前的沟通中，邵总一直比较谦虚，说就自己的背景身份而言，可能不会特别冒进。我觉得这个定位挺好的。认清自己，做适合自己身份的事，对任何人来说都很难。此外，您也希望除了个人建树之外，认清自己的使命，对整个行业起到一点积极的作用。北京院是在国家建筑行业内起领军作用的大型设计院，您是这个平台上的执行总建筑师，同时也是中国建筑协会建筑师分会的理事长。这两个身份定位锁定了您的使命，逼迫您更要认识自己：已经不能从个人小趣味、小情调出发，而要从更宏观的角度引申出学术意义和社会意义。另外，您也希望不辜负设计生涯中获得的设计机会——与库哈斯、福斯特、扎哈、盖里这些国外优秀团队合作的机会，也对您的职业认知有自觉和不自觉的影响。

邵韦平　　我去年和盖里在一起工作过，我觉得像他那种风格的建筑师都具有极强的独特性，模仿根本就没有意义，因为那种思维方式绝对是只属于他个人的。所以结合本人的成长背景，我认为我们的工作方向，不会像盖里、扎哈、库哈斯那样完全靠个性化风格取胜，我们可以做另外一类建筑师，比如像努维尔，可能更灵活一些，根据特定的项目功能、场所条件做特定的高品质建筑，每个作品不一定追求单一的风格。可能大部分的职业建筑师最终都要走这条道路，因为并不是每位建筑师都有条件按单一的标签从事职业工作。大量的高水平设计合作，让我们看到了自己的不足，也为设计团队提供了宝贵的学习机会。

王辉　　但都不是浅尝辄止，而是比较深层次的、职业上的交流。这跟一个简单的来往，或从书本上读来的经验很不一样。

邵韦平　　通过这种合作确实有一些比较深层次的专业心得。

王辉　　是，我听您说过好几次。我觉得您在这方面是比较敏感的。我相信，有过这种经历之后，您会对自己的职业有所触动和反思。这些机遇的积累，当然还包括您自主设计项目中各种摸爬滚打经验的积累，都会对工作室的定位有决定性作用。它变成一种因果关系，同时也决定了工作室做什么、不做什么、怎么做。正是因为您在单位里的地位、社会地位、职业地位等一系列条件，有意无意地获得了不同一般的职业经验和阅历，对工作室的轨迹形成了一个因果关系，而不是随性关系。

| 邵韦平 | 不过这种大院背景也会对我们形成一些体制上的制约，有时候我也担心，由于大院的惯性使我们的工作失去前进的方向，比如被一些非专业因素左右。在大院里克服这种惯性是要有毅力的。受分配体制的影响，与其他团队配合的时候，做简单的设计容易，做复杂的设计对缺乏追求的设计师来说就成了一种负担。我们也很羡慕都市实践这样的团队，更自由，工作方向更符合自己的专业价值。不过反过来想，北京院也有北京院的优势，专业技术储备多，有坚实的基础，做一些探索也撑得起，不至于轻易就破产了。但如果自己控制不好的话，往往会流于平庸，这是我们体制里容易犯错的地方。所以，我一直努力倡导一种创作氛围，哪怕存在失败的风险，也要努力尝试一下。做凤凰的时候就是这么想的，设计刚开始时都没想到能达到现在这个效果。 |

| 王辉 | 凤凰这个项目开启了建筑技术和设计方法上很多新的东西。不能说先河，我也不知道用什么话或者词汇来形容，但应该是开拓了建筑学在技术上、理论方法上的窗口。我比较佩服邵总愿意花精力来投资，将团队交的学费转化成教材，让社会分享。 |

| 邵韦平 | 一个具有创新意义的设计一定需要开展深入的探索研究，研究环境、人的行为模式以及可能的技术构成。凤凰的创作过程就是一个研发过程，我们投入了大量的精力和财力，为设计定制专门的技术体系和实现方法，而这些技术体系和实现方法比建筑呈现的表面形式更有专业价值，我们也十分看重它的意义。因此我们团队的核心人员一直在集中精力，整理和归纳过去积累的海量技术素材，希望能尽早完成，满足大家的期待。 |

| 王辉 | 我都觉得自己老了，邵总是比我高好多届的学长，还能勇于接受并运用数字化新技术，我很佩服。这幢建筑确实在很大程度上推动了行业的发展。很多人可能会觉得门槛很高，这些东西是不可企及的。这幢建筑是一个完全原创的自主设计项目，完全在中国技术与中国智慧的条件下成功的典型案例。我相信您的工作室人才辈出，人人都是各方面高手，但我也相信他们只是普通的凡人。在这种组合下能做成这种事情，其中蕴含的经验是很有价值、值得分享的东西。 |

| 邵韦平 | 在凤凰这个项目上，我们正在做的是把已经完成的数字模型与运行维护结合起来，比如灯光、空调和消防控制等，进行三维的仿真运行监控，这样模型的作用就会越来越大。我们在做新的项目时，也开始将数字手段和建造安装、运行维护结合起来。这种全生命周期的设计思维还是很有前途的，将对建筑的品质起到重要的促进作用。但是从现在行业的实际情况看，包括我们设计院，应该说数字模型运用还处在一个学习探索的过程当中，要达到理想的境界还有很长的路要走。 |

王辉　　　　有时候我会这么想：北京院的一个所就能做成这件事情，可见门槛不一定那么高，只要努力，这扇门就能打开。所以这件事如果能在社会上得到普及，对很多人来说都是很大的鼓舞。邵总愿意把这些经验放在公共平台上，跟大家分享，在普适性的推广上意义更大。

说回到自由。我觉得不自由是绝对的，自由是相对的。我们都受到各种因素的约束，做大事就会面临更大的不自由。换句话讲，这种更大的不自由就是对自己要去做的事情的怀疑。我也是这样，每做一件事情都在想它的意义是什么。但是，也就是在超越这种不自由的过程中，我们获得了一点动力和享受。您说你们面临的问题，其实我们面临的问题也很多，幸福的家庭都一样，不幸的家庭各有各的不幸。每个人在自己的位置上、角色里，都会碰到过不去的门槛。您能到今天这个位置上，肯定也是久经考验，非常有领导力。

邵韦平　　　　不是一个人在战斗，我们的团队集聚了一批经过复杂工程磨炼，仍然坚守下来的优秀员工，他们出色的表现是工作室取得进步的必要条件。

王辉　　　　从社会风气来讲，人们对于愿意探索、愿意付出的人还是肯定的，像咱们这种量级上不匹配的设计单位之间，也互相欣赏，互相支持。这还是一个做事情的时代，在这种情况下，我们这种自我奋斗型的事务所和你们这种带有任务式的代表大院水准的设计部门都很务实。事情做多了，就是另一番天地了。像努维尔跟北京院合作，我看过他的设计方案，一旦建成了，还是能促进材料工业、建筑规范的发展的。你们与他们配合，对自我发展也有好处，这是一个良性循环的过程，也是克服不自由的过程。

姜珺 × UFo

姜珺，研究型建筑师、文献编辑与自由撰稿人，《城市中国》杂志创刊主编（2005—2010），2014年威尼斯双年展中国馆总策展人

Jiang Jun, Researcher type architect, Literature editor and Freelance writer. Founding editor-in-chief of *Urban China* magazine (2005-2010). Chief curator of Chinese Pavilion in 2014 Venice Biennale of Architecture

3.3

李涵　一直以来，UFo关注自己的执业状况在国内、国际范围的影响，也乐于参与行业内外的媒体活动。那么，先从你前段时间参加的展览开始谈起吧，关于中国设计机构的展览。

姜珺　那个展览叫"东方愿景"（Eastern Promises），由维也纳的MAK美术馆主办，但策展人是独立于美术馆的，主要关注东亚的建筑现状，覆盖了中日韩几个主要国家。展名中的"愿景"概念，是希望通过这些东亚参展者的作品和研究展现东方人的理想空间。参展人在取向上选择各不相同，我这部分是一份关于中国设计院的研究性文字。

李涵　虽然设计院的工作占据了国内绝大部分的市场份额，但这类独立展览一般对设计院关注不太多。你这次参展的研究是仅专注于大设计院这个圈子吗，而不是对整个建筑界的研究？

姜珺　基本上以设计院为重心，但以建筑界作为其语境。比如，1950年代设计院体系初创时，人员大都是民国时期独立执业的建筑师，那么设计组织的变化对他们的建筑思想有何影响；1980年代设计权下放，设计院在面对海外事务所和民营事务所的竞争时又有何变化等。展览里多处涉及北京院，包括北京院初期和苏联的合作，当代北京院作为"现代国营企业"的机构图解，以及像UFo这样的直接面对市场的品牌工作室在北京院传统框架中所起的作用等。我接下来在威尼斯双年展的中国馆策划中可能也会涉及这方面的内容。

设计院体系是一种设计资源的配置手段，最早作为苏联模式的一部分引入中国。国共两党在首都建设上都遇到过相似的问题，就是要集中资源搞建设，首都的任务就是要纳入一个具有集权特征的现代民族国家的党政军机构，大集中小分散。这就需要有设计力量来协助。1952年到1956年间，设计院体系作为社会主义改造的一部分形成，同时作为资源统筹配置的一部分。1958年中国第二个五年计划遭遇"苏援"紧缩，于是有了"大跃进"，一方面动员地方二级政府，以区别于苏联高度集中的计划经济；另一方面以中国过剩的人力资源替代稀缺的资本，所以"大跃进"的结果不是私营事务所的复活，而是大量地方设计院的产生。

设计院这条线索和中国的地缘政治、经济模式高度相关，这其中有利有弊：利，就是在冷战时期的大规模建设中，能够以快速高效的方式实施国家战略；弊，就是所有托拉斯大企业共有的弊端，机构臃肿，制度僵化，活性低下。总的说来，设计院这样的大组织有利于技术层面的集体协作，但不利于文化层面的个性发挥。

| 李涛 | 所以你的研究主要是从1949年一直到现在？ |

姜珺　是的，之后的一个时间节点是1984年。当时中国出现了第一个私营建筑工作室，尽管作为部院下挂的工作室，其所有权还是国有的，但经营权已经下放给个人了，财务也是独立的。这一制度改革和当时发生在农村和沿海特区的改革大致平行。邓小平时代改革的特征是摸着石头过河，不是一步到位的休克式疗法，本质是把毛泽东时代集中起来的资源作为红利，分期放权给社会，每一次下放都会有新活力，也出现新矛盾。被设计院垄断的设计权也在这一时期下放，但主要还是在设计院体系的框架下，比如为了解决农村剩余劳动力的就业问题，建筑市场对农民工开放，当时就出现了低门槛的丙丁级设计院。

到了90年代，农村市场让位给城市市场，农民包工头做设计的少了，但学计算机的、学工程的、学舞美的都开始做设计了，这个行业在快速城市化的轨道上就是个暴利行业。资质制度是中国上万家设计院得以存活至今的原因，2000年之后大量境外事务所来华实践也需要设计院的合作。不过设计院也有它的优势。大量的建筑规范都是本土规范，很多大项目的落地，需要设计院尤其是大院在过去半个世纪中的经验积累。大院有它的综合素质，不光是机构间的协同，结构照明水暖电与设计的协同，还有专门的研究部门，国外的设计机构里面很少有这种情况。

李涛　刚才提到国有大机构创造性的问题。社会上确实有这样的观点，认为体制内的设计院把持着资源优势和庞大的市场，但没有发挥出更大的创造性去影响城市，没有对这个行业负起更大的责任。从你的研究中，你认为这一现象是什么因素导致的呢？

姜珺　一是国有制度在激励机制上的问题；二是垄断问题，金饭碗里养不出创造性；三是机构大到一定程度就必然出现效率低下问题。现在很多大院都有内部竞争机制，但其结果取决于上级而不是社会评价，很多大院的个性取决于上级的个性。所以某种程度上，其创造性也取决于其上级的视野和包容程度。

李涛　你说的情况确实存在。所谓大院个性也不仅仅取决于上级的个性，从大的范围来看，作为国企，大型设计院的各个方面，包括创造活力，必然由上至下地随国家的活力、开放程度有所起伏。近些年，大型设计院内部也出现了少数活跃的具有相对独立价值观的设计团队，他们的实践能否由下至上地回馈、影响整个大企业的价值观，进而改善整体创造活力，还有待这些团队和整个企业的思考和努力。

姜珺　设计院体系创立时强调共性大于个性，这是国家资本在原始积累时期的历史现象。但在今天的设计环境中，再固守不变就无异于刻舟求剑。中国改革的特征之一就是倒逼，一是自下而上，一是由外而内。设计院体系就是在一个逐渐由市场主导的设计环境中，面对来自民营和海外团队的倒逼；在设计院内鼓励 UFo 这样相对独立的设计团队在方案创作层面参与市场竞争，也是这一倒逼的结果。正因如此，我才认为需要有对中国设计院体系的系统化梳理，了解自身和环境的关系和走势，知己知彼才是在一个良性竞争环境中的生存之道。

刘宇光　有没有听过国外的机构对中国设计院的看法？似乎中国设计院近些年来一直作为 LDI（Location Design Institute 本地配合设计院）的身份出现，其实也有一些设计院的建筑师有个人或团队的品牌，活跃程度逐渐升高，但在国际上的知名度和影响力还有欠缺，因为设计院的建筑师更多关注工程技术层面的问题，对媒体和传播的关注相对比较少。

姜珺　很多人都不了解中国设计院，这不奇怪，就连大部分国人自己也不了解。这次 MAK 的展览，在我阐述概念之前，策展人几乎对此闻所未闻，但他们有很好的直觉，设计院体系在中国有这么大的潜在影响却不为外界所知，就一定要纳入他们的展览。有人认为设计院不出作品就乏善可陈，这是一种脱离历史的观点；另一些了解历史背景的，又可能会带着意识形态的有色眼镜来看。

刘宇光　在近现代的建筑史当中，俄罗斯和一些东欧国家，比如匈牙利、捷克、克罗地亚和斯洛文尼亚，它们本来有长期的建筑文化传统，一直都有很多建筑大师，但自从采用苏联式的设计院体制后，一直到今天，再也没有出现过有国际影响力的名字。这是因为它们自身水平下降了，还是不跟国际社会往来，失去话语权了？在今天，建筑师本身也是国际市场经济的实践者，如果不掌握市场的规则，也很难找到施展的空间。

姜珺　1930 年代之后，构成主义的主角们就像宋江一样被收编了，不愿意被收编的也被边缘化了。新古典主义是近代大部分国家威权的共同选择，所以赫鲁晓夫时期对斯大林威权的批判在建筑上是有解放意义的，他通过标准化建造为更多老百姓提供了居所，尽管按照现在俄罗斯官员的说法，他建的房子就像地狱一样，不过要考虑到当时他追求的是经济性和社会性。苏联时期还是有不错的作品，只不过不为外界所知，比如在苏联被叫作"米拉克杨"（Microrayon）的小区，也就是具有小社会特征的"微型城区"，还是有其先进的一面。反而是在苏联解体，同时也是苏联设计院体系解体之后，真没有什么有影响的建筑师了。这里有后苏联时期人才流失、工程腐败、建筑教育模式等问题，但总的看来，俄罗斯文明的黄金时代似乎是接近尾声了，这还不是俄罗斯建筑界独有的现象。

刘宇光　在中国，像张镈那一代建筑师，体验了一次建国立业的过程，他们是从私营事务所的工作状态进入国营大设计院的工作状态的，社会政治文化背景都变了，对设计本身一定会产生很大影响。

姜珺　　　　　新中国成立前他们很难做大项目，新中国成立后的项目都是中国最大的项目，像三里河行政中心就是超大型项目。很多人把注意力集中在大屋顶去留问题的辩论上，但我认为这个方案更重要的是它以分形宅院建立起的一种大院模式，可惜只盖了1/9，但即便如此也能供"四部一会"使用，可见这个项目有多大。而当时的张开济才40岁出头。北京院一个更好的例子应该是张镈，新中国成立前他所在的基泰工程司号称近代中国最大的建筑事务所，这个事务所在1930年代首都计划和大上海计划中有大量参与，但主要还是因为其创办人关颂声和国民党上层的裙带关系。新中国成立后像部院和北京院接手的几乎都是重要项目，但也因为大项目的政治性质，建筑师如履薄冰，北京院的"二张"都是当年大屋顶批判中的首当其冲者。相比之下，国民政府时期也有过对民族形式和经济性的辩论，像范文照，后期对自己早年的民族样式做法是有反思的，这恰恰说明那一时期建筑师还有自主选择，这和新中国成立后设计院体系中的建筑师状态就有很大区别。

刘宇光　　　　北京院著名的"八大总"，当时就是一支国家队啊。他们当中有些人既是建筑师也是规划管理者，拥有很大的发言权，社会影响力要比现在高。

姜珺　　　　　是的。不过话说回来，在现在这种什么都不确定的情况下，到底是由领导主导，还是由专家集体主导，还真不好下定论。有时候领导的看法真比专家看得要远，专家作为一个固定职业的从业者，有时会当事者迷。建筑师这个职业最忌宅在院里闭门造车，还是需要经常输血，才能保持更新。比如说像你们这种编书、邀请专家做院内讲座、参与国际竞赛等，把自己当作市场里一个能动的个体，能够跟外部的新鲜血液进行常态的沟通，我觉得是有好处的，至少可以拿来作参考。

李淦　　　　　UFo在这方面还是比较开放的，外界对我们的印象也是如此：一方面我们作为北京院的机构，具有大院的技术优势；另一方面，我们也是大院背景下有相对独立价值观的一个团队。对公众性的活动一直保持着关注和参与，能发出自己的声音。就像几年前你给邵总做的访谈起的题目——"大院里面的UFo"。上一次设计院的研究对大院将来的发展有什么探讨吗？

姜珺　　　　　设计院体系的改革和中国国有企业的改革是平行的。目前国家的态度是让掌握非核心资源的国有企业参与市场竞争，设计院就属于这一类。北京院就是这样一个观察对象，我在文章中为北京院做了一份组织结构图，讲决策和分权的关系。国企改革的大方向，还是在国家主导的前提下进一步社会化，国资委的资金是全民所有制而不是国家所有制，当然现在还不现实，还是要国家代理我们管理这些钱，以国企收入作为国家财政收入的一部分，因而也间接地降低了税收的压力。随着私营建筑师事务所制度的日臻完善，国有设计院将逐渐丧失其垄断性，作为市场主体的国有设计院如何在国资委主导下纳入现代企业的活性，绩效和收益挂钩，资本和管理分离，脱离行政干预设计的家长模式，将是大院改革面临的主要命题。

尚世睿 × UFo

尚世睿，时任SOHO中国副总裁，分管北京的项目设计管理及研发、工程信息化协同管理工作；负责北京银河SOHO设计管理及项目管理工作，以及望京SOHO的设计管理工作；耶鲁大学建筑学学士、硕士，清华大学建筑学学士，纽约大学房地产金融硕士

Shang Shirui, Former Vice President of SOHO CHINA, in charge of project design management & research, and informatization cooperative management of projects in Beijing. He directed the design and project management of Galaxy SOHO, and design management of Wangjing SOHO. B.Arch and M.Arch of Yale University; B.Arch of Tsinghua University; Master of Real Estate Finance of New York University

3.4

李淦

我们接触过很多不错的甲方，但像SOHO中国这样的甲方还是比较特殊的：一是控制比较精确，在设计的各个阶段，无论是方案还是施工图，在出图前相应部门对设计文件都有详细的审核，甚至对图面都会提出意见；二是设计一旦定下来就不会轻易改变，这样执行起来也比较顺畅。虽然SOHO的项目通常规模大且复杂，但执行效率非常高，建设周期也比较短。

尚世睿

分析得挺对。SOHO做开发有将近20年的时间了，银河SOHO和三里屯SOHO这两个项目可以说是它本身产品线的升级。如果朱院长设计的现代城SOHO算一期的话，之后的建外SOHO已经开始了对城市的思考。尽管外界对建外SOHO有很多指摘，但不得不承认它在城市方面是相当成功的，融入了市民活动的空间肌理中。到尚都SOHO和朝外SOHO的时候，SOHO实际上是在城市设计不断完善的情况下，琢磨怎么做建筑本身，这时候有各种尝试，最后到三里屯SOHO和银河SOHO有了质的飞跃。而SOHO的设计管理也伴随着这些项目在成长。在银河SOHO设计的过程中，我跟李工也探讨过这个问题：在中国，甲方之所以需要投入这么大的力量去做管理，实际上是设计总包的缺位造成的。

银河SOHO能最终严格控制住成本，是建立在三里屯SOHO项目经验基础上的。由于当时的一些项目进度因素，三里屯SOHO在结构形式、机电图纸上，由两个设计院反复调整，设计总包的缺位导致了设计管理一定程度上的混乱。

我们和扎哈也不是第一次合作，早在朝外SOHO的时候就请她做过方案，后来她也参与过物流港的设计。到银河这次合作的时候，我们也比较懂得怎么跟外方设计师合作了。一定要做到信息实时共享，从而避免犯大错。同时北京院，尤其是UFo工作室对复杂设计的理解，以及设计深化的能力也是项目能够顺利推进的一个重要原因。我们也合作过一些设计院，你们有能力同时承接凤凰传媒和银河SOHO，是一件挺不可思议的事情。

李淦

我们是一个原创的团队。正是因为凤凰这个原创项目，使我们有了操作非线性建筑的经验，对于其后理解和实现扎哈银河SOHO的建筑形态就比较容易一些。加上北京院和我们工作室多年技术的积累，也是支持这一类项目的强大后盾。

尚世睿

很遗憾你们现在不太愿意做LDI了。

邵韦平

这主要是受时间的限制。我们能参与银河也是一种缘分，在一个很巧合的时间段，有机会与SOHO合作银河项目,对我们工作室而言也是很重要的事件。我觉得这里面有两个值得探讨的地方:一是SOHO中国团队是一个很成功的、控制力很强的团队,在开发方面让我们受益匪浅;再就是SOHO中国选择了扎哈这样一位极具时尚性和挑战性的外方设计师,让我们在做凤凰的同时,能够有和这类外国大牌近距离沟通的机会,对我们也是一次难得的经历。

UFo的大背景是一个传统体制下的大院,我本人是单位的总建筑师,在工作室中聚集了一批很有追求的年轻建筑师。银河SOHO这个项目是一个很好的锻炼机会,在此也希望听听你的建议,像我们这样的团队如何与市场接轨?应该从哪些方面再提高一下,以更加适应像你们这种成熟开发商的需求?

尚世容

我认为,国内设计的概念实际上不比国外差。现代主义的建筑设计,可以从方案质量和设计质量两个部分去衡量。比如贝聿铭的设计,方案朴实,设计质量极高。现在回过头看我们的合作,恐怕UFo这样能提供基于二维协同的设计,在市场上也是为数不多的。

其实我一直不太赞同LDI的工作跟创新不搭边的说法。我在纽约工作时,看到就算是KPF和SOM,他们有大量的LDI的配合工作。LDI的意思是Local Design Institute,也就是说在这个地盘,你对这里最熟悉。在纽约干活,一定是SOM、KPF等这种公司"本土"公司最熟悉当地的法规,最熟悉当地的各种人文条件。比如伦佐·皮亚诺要盖纽约时代(New York Times)就一定要咨询这些大公司,而这些公司对他的设计的修改不能说没有原创性。就我们合作过的几个项目来讲,北京市场上的北京院、上海市场的同济院和华东院无论从服务理念上,还是从服务的技术上来讲,都起到了同样的作用。

邵韦平

我对你的观点特别认可,最近我也在思考这个事情。我觉得建筑设计不能停留在一个特别模糊的概念阶段。中国建筑师很容易出现这个问题,就是把工作停留在早期特别粗浅的阶段,之后戛然而止,接下来就交给分包单位和合作方去做。这是一个缺陷。最近我看到一些早期现代主义大师的观点,我觉得在中国对现代主义建筑没有形成一个准确的判断,很早就将其定义为少就是多,或者功能就是一切、形式追求功能等。所以,我们这一代或者更早一代人所接受的建筑教育,受时代的影响,把设计变成理科类的技术性问题,并没有深入探讨下去。柯布西耶曾经说过,建筑设计包括创意和落实创意,需要把创意构造出来。我们从这一点延伸思考:一个当地的设计师,在合作项目里到底应该做什么?很多设计团队认为,合作设计就是帮外方审一审规范,填填图签,把结构、设备、电气对付上就可以了。其实这样很不够。跟SOHO配合的时候,你们就说过结构设计师可以参与方案的优化工作,这种概念在以前的国内设计院设计过程中,还是很少涉及的。我们做SOHO的时候就明显体会到,设计师需要关注整个体系,包括一些技术环节。

要做一个好作品,除了要有好的创意,还要有把创意变成现实的具体措施,我觉得这是现代建筑发展的必然趋势。像建筑总包的问题,我认为中国现代设计中有一些观点是错误的,比如说装修的问题,幕墙的问题,建筑师往往以甲方没有给我付费为由不参与相关工作,最后造成设计深度不够,因为这个设计不是建筑师做的,而是厂家做的,厂家跟建筑师的觉悟完全不一样。而且,长时间不介入这类专项设计,建筑师也不会做。所以我认为,现代的核心矛盾不在于甲方有没有给钱让我去做,而在于我们到底想要什么,是想要在建筑师控制下的作品,还是追求利益最大化,把它当成商业手段而已。在国内经常出现这个问题,本来想做好东西,后来又把初衷撇到一边,去追求成本控制。

北京院也有这个问题,有些生产部门以成本定标准,造成深度下不去。无论是做银河SOHO还是做凤凰,都是超越了常规的。不过我们也要过日子,也要有必要的经济回报。也许当我们有了这种能力,获得更多信任之后,才能够得到回报,这是一种良性循环。有时候甲方给予了信任,设计单位却承担不了这个工作,这是特别可悲的事情。要做一个好工程,还需要优秀的业主,能够和建筑师默契地配合。

李淀 说到设计总包,可能跟甲方的设计管理很有关系。通常我们院接的任务,设计范围仅仅是土建设计,不包括室内、景观的设计。而SOHO中国委托建筑师的工作,就全面包括建筑、室内、景观。

尚世睿 你说的特别对,这是一个鸡和蛋谁先存在的问题,到底是你先把能力拿出来,还是我先把合同范围给你?在国内面试设计院的团队时候,设计师需要一一介绍在某个作品中所做的工作,很多时候解释下来,一个项目这里是外方设计的,那块是幕墙公司做的,室内有专门的室内设计公司,最后就剩下一个混凝土的壳,再加点机电是自主设计的。这次和UFo的合作是一次深度合作,实际上打破了上面的困境。因为北京院本身是一个改制的单位,单个工作室没有机电,没有结构,需要在北京院内部"跨单位"协调结构、机电资源……比如我们从各个方面综合考虑,要求办公室的室内是裸顶,不吊顶,那么机电设计就不是一个简单的线路不碰撞的问题,而是需要设计,像蓬皮杜中心一样,机电管道本身就是立面。所以这种情况下,MEP(机电设计)就是重中之重。正是因为UFo有胆识,综合能力强,使得沈工和杨工能够充分发挥出潜力。("沈工"为银河SOHO设计团队设备专业负责人沈逸赟,"杨工"为结构专业负责人杨洁)

邵韦平 我们的结构团队非常优秀,只是不太善于表达,他们是能够挑战一切极限的。

李淀 束总(BIAD结构总工程师束伟农)是银河项目的结构负责人,有次杨工和他讨论结构方案,说到现在扎哈方案中的连桥平面是弧形的,这种结构咱们以前没有做过,束总回答:"没做过就做一个吧。"

尚世睿	（笑）这句话是有狠劲的，一听就知道是愿意创新的设计师。

| 邵韦平 | 关于建筑师和专项顾问的关系，我认为建筑师一定要积极参与一些重要的专项设计工作，不然最终成果往往是一个瘸腿的、不完整的建筑。同时，为了做出好设计，还是要请专项顾问来支持。这一点国外做的比较好，某种程度上也反映出国内的短板。有时候国内建筑师有一点无知者无畏，像幕墙、装修、景观这类设计工作，明知很复杂也不找顾问，随便画一个出来。当然我们自己的工程也可能有这个问题，有时候也是机制造成的。我们与KPF合做Z15中国尊的时候，有一些顾问节点控制非常严格，他们会督促甲方做决定，很好地保障了设计的正常推进。由此也可以看出国内整体设计运行机制的不完整性。有时候可能甲方不太懂，但是设计师也无所谓，好坏不管，硬做就是了。所以我们后来总结，在一些重要的节点，寻求必要的专项支持是特别重要的，不能因为没有钱，或者甲方没做规定就可以放松要求。刚才讨论的结构、MEP等专项设计，对于保证建筑品质极其重要，而且很多时候这些是进行创造的机会，若是做好了，它本身就是一个艺术品。我们以往在做基本配合的时候，只是想着要把路走通，没有形成一种逻辑、一种秩序。通过做银河SOHO和凤凰两个项目，我们体会到专业设计在品质上的作用。 |

| 尚世睿 | 凤凰的雨水管很多都没法藏起来。 |

| 邵韦平 | 对，结构是完全暴露的，所以要同时具有支撑作用和美学作用。这也是一种设计方法，或者说习惯。开始费了一点时间，但是养成习惯之后，花费的时间并不是呈几何层级增长的，而是可控的。如果不这样做，就要花更大的代价去填补这个空白。比如后期再通过装修包起来，在这个过程中，设计可能就失控了。所以我觉得养成良好的习惯可以提高效率，而且更易达到质量标准。 |

| 尚世睿 | UFo的意识非常好，是真的觉得每一处细节都需要设计控制，这一点在与李工（李淦）配合的时候能感觉到。 |

| 邵韦平 | 一个严谨、成熟的开发商对于设计品质的控制，是很有帮助的。我们的团队成员也很有感触，相比不太专业的甲方，银河项目虽然受业主限制比较多，但工作比较有成效。好的业主能让我们自身的设计机制更加成熟。 |

| 尚世睿 | 下一步就是怎么能够产业化。 |

| 邵韦平 | 让大家都能够掌握这种标准。 |

| 尚世睿 | 让刚进来的新鲜血液也能掌握，特别是在你们人员的流动性还比较大的情况下。 |

| 邵韦平 | 总的来说，北京院，尤其是我们这个团队的市场吸引力还是比较大的。新招来的学生素质不错，学习能力很强，多 |

数很快就能上手。我觉得培养一个有基础的年轻人，比教一个养成习惯的老手容易一些。很多老同志有惯性，设计定得很快，当你发现不好的时候已经来不及了。国外很多事务所一般有一套机制，保证所有的人员一进来就能够统一价值观。这是值得我们学习的。

尚世睿

你们成长的过程就是在消灭甲方的设计部。在美国几十亿美元的项目，业主方也没有一个像我们这样200多人的开发团队；他们完全依赖建筑师、总包及项目管理公司带领的项目团队，所以建筑师合同的含金量很高，这是真正意义上的设计总包，以总建筑师为首。可能还有一个概念建筑师带领众多的顾问，形成一个团队，这些团队的最终服务成果在竣工时由总建筑师验收。而国内的四方验收则沦为一个质量安全的检查，对于设计质量，没人会在这个会上提太多。所以我们业主才会把这个部分的考量融入日常的设计巡检，并产生了大量的巡检报告。以前只有老板看到有问题，才会出巡检报告，现在这个习惯保留下来，沿用到SOHO其他的设计项目，每周会根据巡视报告和项目部开会，讨论整改措施。通过这个办法，我们把工期和施工质量控制在一定的可接受范围内。在这个过程中，扎哈团队和UFo对这项工作都付出了极大的热情。这也是SOHO对于外方设计、设计院的要求，除了满足基本的现场配合的要求，还需要高强度的现场巡视工作。

刘宇光

我想回到另一个话题。你在最开始强调了SOHO中国在城市设计方面的成功，从建外SOHO到三里屯SOHO，每一个项目都成为城市的活力区。对于项目最后达成的效果，你们是从制订任务书开始，就要求建筑师朝这个方向去做，还是在选择最终方案的时候朝这个方向去靠近？我很想知道你们的引导方式是怎么样的。

另外，你们在逐步形成的城市区域中，通过什么方式来培养人们的生活方式？SOHO这样的模式在美国的一些大城市和东京很成功，因为当地人的素质很高，居民之间互动营造的公共生活很成功。但是在中国，人们的整体素质可能达不到这种需求，所以结果往往背离了你们原本的业态设定。那么你们在成长的过程中，有没有考虑过将来业态和城市生活的组成方式呢？

尚世睿

这个话题大了（笑）。我没法代表SOHO中国回答这个问题，只能通过这几年的观察，给出一些个人观点。首先，SOHO中国对于公司产品非常重视，在选择设计师方面乐意启用明星建筑师。李工曾问我，为什么SOHO中国不请设计院做整个设计，像UFo这样的团队也有很强的原创能力啊。事实上，SOHO中国有一种偏执，认为大事务所，无论国内国外，首先关注的是经济利益，这在一定程度上会减弱其原创动力。有一些事务所的状况也确实印证了这种想法。另一方面，SOHO这几年积累了不少散售型商业模式的经验，对于这种模式相关的设计需求也在逐步清晰。至于散售模式可能跟小业主的素质没有太大关系，用素质这个词形容可能太严肃了（笑）。我认为SOHO更崇尚自由生长。

王舒展 × UFo

王舒展,《AC建筑创作》杂志社主编,高级建筑师,国家一级注册建筑师,清华大学建筑学硕士;京沪高铁南京南站工程建筑专业负责人,北京银河SOHO设计总负责人之一

Wang Shuzhan, Chief editor of *ArchiCreation*; Senior architect; National first-class registered architect; M.Arch of Tsinghua University. Architectural director of Nanjing South Railway Station project along the Beijing-Shanghai high-speed railway; one of the chief directors of Galaxy SOHO project

3.5

李涅 　先聊聊建筑设计机构和媒体的关系。几年前我们做过一个关于设计机构知名度的分析研究，用谷歌搜索一些设计机构的名称，从搜索到的条目数量上看，国内几位明星建筑师在30、40万级，咱们北京院在300万～400万，OMA达到了7 000万级。从条目内容来看，国内明星建筑师和OMA这类机构有一个共同特点：他们不只出现在建筑领域，在时尚界、艺术界和文化界都有涉及。而北京院差不多只局限在建筑专业媒体。可以说，北京院面对的或者说介入的媒体领域一直比较窄，几乎不涉及大众媒体。

2007年，UFo参加深港双城双年展。令我印象深刻的是，在这一重要的展览中，我们是唯一发声的国有机构团队。之后UFo参加展览、活动的频率和级别也更高了，比如像威尼斯双年展这样的顶级展览。

王主编曾经是职业建筑师，而后从事建筑媒体行业，关于设计机构和媒体的关系一定有很多体会。很想请你聊一聊你观察到的情况，北京院、院外机构再扩展到国外的建筑机构，他们参与专业媒体、大众媒体的状态如何？

王舒展 　这两年《AC建筑创作》杂志社接触了不少事务所，从国外大牌、老牌设计师，到国内不同规模的设计企业都有。我的感觉是，国外的企业，不论性质、规模、文化的差异，在对待媒体这方面与中国企业相比确实有很大不同。首先，国外企业不论规模大小，都有一个专门的部门来对接媒体，这个部门也许很小，只有一个人，但是一定是有一个专职岗位的；企业向外界透露的一切信息都要经过这个部门，发布信息的权限在企业内部被定义得非常清晰。其次，这个部门或专职岗位不只是一个对外的窗口，它的存在有更大的作用在于，它的职责首先是将公司的项目、业绩进行有序地整理、分类归档，包括项目演变过程中的图纸资料、最终的图纸文件、建成作品图像信息，等等。每一个国外企业都有这样一个丰富有序的资料库，在此基础上，与媒体对接时，经过双方的努力，就能比较高效地对资料内容进行升华，得到一份经过某种包装的、可读的、易于被人接受的媒体产品。由于这样的基础工作比较扎实，媒体产品的调性与水准可以得到基本保证，不管面对什么类型的媒体输出，都能够树立较为统一的企业形象。而这点对于大公司尤其重要。

在国内，在媒体上往往能看到一个大型企业的很多个侧面，有的可能很高端，能与国际接轨，有的可能就是国内一般水平，你不知道哪一个形象能真正代表这个企业。但是国外企业不论内部状况如何，对外的形象一定是统一的，同时也传递一种统一的价值观和企业文化气息。前几天我和一个学传播学的朋友聊天，他向我解释了原因：要想把信息有效地传递到人的头脑中，统一的信号非常重要；否则，就算信息再丰富，也给人留不下什么印象。

我们这种国有企业，现在也在变化，北京院在这几个企业中算是走得比较靠前的，比如今年北京院的网站可能要改版了。但是这个事情还在酝酿，我们也在参与出方案和咨询报告。

李淬　　　　　当年我们工作室也做过一轮北京院网站的界面设计，但是没有被采用，可能是因为过于强化北京院作为设计机构的特质，与院里定位不太符合。看来现在还是要做一些改变？

王舒展　　是的，但是我觉得从酝酿到最后达成结果，路还是挺长的，估计最终做决策还是不容易，因为这关系到怎么定企业的调性。但是不管怎么样，现在要比前几年条件成熟了，因为我们真正转企了，现在做这个网站的目的非常单纯，就是要对外宣传，赢得我们的客户。所以我觉得在往好的方向发展了。

李淬　　　　　上次参加威尼斯双年展，我们已经觉得自己的投入比较大了，出去一看，外方参展投入更加惊人。

邵韦平　　这是文化的差异。国内在建筑文化方面投入不足，因此在国际展场上，我们显得很孤单。这反映了一种观念上的差距，在这个方面，中国还有很长的路要走。我们也需要不断总结，正如你前面说的，在作品的整理、开发方面，我们跟国外相比还是有很大差距，因此要利用这次出版的机会，把它们整理出来，拿出与我们的追求相匹配的作品来。

建筑师这个职业有它自身的规律性，虽说人各有志，但还是可以找到正确发展方向的。我特别感谢北京院给了我们这么一个天地，在工作室能过滤掉很多杂事，让我们保持理想，可以实现一些职业梦想。为什么要出版这样一本作品集呢？我们觉得UFo在体制内的环境中，还是挺另类的，这些年也有一些收获。我们将团队的创作热情与大院的优势结合起来，没有满足于一般的设计标准，在创意和新技术方面还是取得了不少成果。

这些年工作室的产值并不是很高，但仍然坚持做这一类项目，也得到了社会上的一些认可。从某种意义上说，在学术方面社会和业界还是挺公平的，像凤凰中心这个项目就完全是靠事实来说话，大家对这个项目的总体评价是很正面的。上次马总在视察之后说，通过这个项目感觉到建筑专业在高科技方面是可以有所作为的，可以和其他专业一样在新技术上有所突破。这样一些经历，更加坚定了我们走品质之路的决心。凤凰走到今天，我们确实为它付出了巨大的代价，但还是挺享受的。因为我们的目标就是这样的，在一个不是很健全的市场环境中，努力走出一条追求品质的个性之路，起码能够对得起我们的职业生命。当然理想和现实总是有距离的，好在我们团队有一批特别有激情和灵气的设计师，团队气氛很好，大家很有凝聚力。经过这些年的积累，我们现在承担的项目大都是这个城市中重要的项目，像中国尊，CBD核心区公共空间等。应该说，这些项目的技术挑战与我们设计师的能力之间还是有距离的，因此我们没有任何可以自满的条件，需要不断弥补这些差距。总结和提高也是我们出书的目的，也希望通过《AC建筑创作》找到一些动力。

王舒展 我们先完善自身，先提高自身水平。（笑）

邵韦平 我经常看到社会上一些设计师缺乏职业追求，设计的建筑不像样子，但是他们觉得已经够了，品质对他们来说是多余的，这是很可悲的事情。我们不甘于做只满足一般需求的平庸工作室，希望有机会跟处于世界前沿水平的同行进行平等的交流，虽然距离成熟的团队还有很多的事情要做，但我们还是有自信的。借这个机会，也想听一下你觉得我们该如何进一步改善自我。

李淦 王舒展的经历比较丰富，之前是职业建筑师，现在转到专业媒体界，成为《AC建筑创作》的主编。我想知道，假如有一天你又回到建筑师这个职业，那么媒体从业经历会给你的思路和决断带来什么样的变化呢？

王舒展 这个挺难的，我只能想到哪说到哪了。UFo已经有了很好的基础，三年前我离开的时候水准已经很高了，现在又有了新的面貌。刚才说到凝聚力，大家有共同的理想，这点特别难得。UFo有一个显著的特点是，工作室很开放，跟世界范围内的建筑设计发展几乎是同步的。工作室的年轻人可以得到各个方面的滋养，一方面是学习到北京院原来在工程方面的坚实的技术落实能力，另一方面可以亲身参与国内外著名设计团队的国际化合作，同时感受到世界最前沿的设计理念与中国最具体的现实国情。两个方向，渠道都非常畅通。我觉得建筑设计团队最重要的工作是人才的培养，而UFo为设计人才的锻炼与成长提供了优越的环境。刚才您说到有的团队劳累忙乱昏天黑地，在基层奋斗的年轻人的生存状态可以感同身受，每天生活在没有"氧气"的环境中，往往对事业甚至人生都很迷茫，没有引领，找不到出口。我们杂志希望提供一种可能，让年轻人知道可以换一种方式活着，可以将我们的职业活出更丰富的层次，更有意义和趣味的方式。UFo在这方面已经走在了前面。

如果说未来提升的目标，我想应该是再进一步梳理理论和体系，修炼内力。我们要形成很结实的后台，从技术基本功、技术管理协同，到建筑设计方法论，再到建筑理论和价值观。高手打拼不是靠招数，都是靠内功。内功的修炼不能一蹴而就，需要每一天的努力。

李淦 你说得特别对。在给这本书做前期策划的时候，我找一些员工开了一次讨论会。大家谈到，UFo算得上是一个品牌团队，但是目前确实还没有形成特别明确的设计理念。所以这本书采取的还是记录的方式，用图解、照片、对话等形式直接地呈现信息。不过从状态来讲，UFo毕竟还是一个相对年轻的团队，不必急于去定义理念。我相信在今后的工作中一定会形成可以激发思考、指导实践的理论和体系。

邵韦平　　　　深思熟虑的理论或者理念可以让行动更有效，避免无谓的冒险与实验。

王舒展　　　　每天进行学习、保证知识信息的输入很重要，不然仅靠原来的积累很容易透支。我自己比较苦恼，每天都在处理各种杂务，感觉在现实生活中，想让自己持续地学习还是挺不容易的一件事情。

李淦　　　　这点可以回到设计机构和公共媒体关系的话题。作为建筑师，最大的责任应该是为公众服务，但是能听到的公众的声音并不是很多。比如我平常接触最多是甲方，不是政府就是开发商，接收的信息比较单一。但是通过与媒体的接触，参加一些展览，或是把作品放在网上，就会从别人的评价中获得一些我们平时得不到的信息，可能会据此调整自己的设计观，也许这也是一种"输入"。举例来说，规划地块北面有一幢住宅，到底要以什么样的立场去看待这个环境？只是达到规范的最低要求，还是给它让出更好的日照环境？也就是说，究竟是从项目利益角度出发，还是从公众的角度去考虑问题？所得出的判断是不一样的，这也是为什么我认为建筑师需要参与公共活动，获得公众反馈，可以调整自己的价值观。

邵韦平　　　　出版也有这样的考虑。通过出版获得交流的机会，总结整理创作的历程。能这样思考也说明大家都在成长。

王舒展　　　　确实，李工刚才说的特别重要，现在只从书本上学习是不够的。

邵韦平　　　　现代建筑理论早年在中国的传播不是很完整，很多理论经过编译之后已经异化，不准确了，建筑专业的学生被很多所谓的主义和口号所迷惑。所以很多时候，建筑教育无法培养出成熟的职业建筑师。我最近看柯布当年描述现代主义的理论书籍，他说的很多思想和人们后来对现代主义的理解是有区别的。柯布的作品很精细，他作过很多探索，想法很有创意，像朗香教堂是一个具有未来感的建筑，但是我们的理论将其描写成了粗野主义，一下子就把柯布扭曲了，没有理解他的真谛。他说设计建筑就是在创造一种意志力，设计的过程就是在寻找一种秩序。我们呢，往往设计都做完了，也没有想到要建立秩序，设计被一些琐碎的事情占据了。我现在认为，做好一个设计就是再创造一种你认为最有价值的秩序，这个秩序要包含并超越任务书和基本功能，这是设计的升华。很多设计做不下去，就是因为它没有经过设计师的精心调理，只是一堆杂烩。没有秩序构架，设计自然也不会成功。

　　　　　　　我们现在做中国尊就有这种体会，做这类建筑必须按与之对应的策略去做，像积木一样搭架子肯定不行。比如结构，超高层有特殊的结构要求，需确认是否为巨构体系是为它而创造的，它超越了简单的楼层板的概念。我们把这个概念扩展开来，先把超高层建筑当成一个系统，这个系统由若干个模块构成，这样就把整幢楼分解成若干的建筑单元。只要把模块之间的关系处理好了，就能控制住整幢建筑。模块之内的事是第二层次。通过这种分级就能保证对设计的控制,在这些方面我们还是有些心得的。再比如说对几何的控制,以前做设计只会用轴线、模数概念,最多再有个尺度的概念,现在我们提出一个广义的几何控制,在美学和功能上就可以做到更清晰的控制。轴线这类概念是对应于建造的,现在要考虑美学、人的感受、完成面、复杂的体形等。凤凰的建筑效果之所以能实现,就是因为我们建立了一套符合它需求的几何控制体系。最开始想找奥雅纳(Arup)为我们提供这套系统咨询服务,但没有合作成,现在回过头来看,我们自己都完成了。

中国设计师是有潜力的,只要有好的环境,就有机会打造出高品质的建筑作品。现在需要转变观念,改变一些落后的设计观念;传统不是简单的继承,还是要用当代人的眼光,从传统文化中提取那些适合现代生活的内容,并把它升华。从这个意义上来讲,我们在创造中国文化的时候,没有必要简单拷贝一些所谓传统的理念和符号,完全可以用更现代的方式去诠释。中国建筑师应该更自信。建筑设计跟电影或其他艺术门类相似,中国人在做设计方面还是有优势的,建筑师在冲破陈旧观念的束缚后,一定能创作出更多符合时代标准的优秀建筑。

王舒展 不管是做建筑,还是做书、做杂志,保持您这样的想法的,还是少数派。做少数派需要勇气。凤凰中心项目完成度很高,我能想像大家为此吃了多少苦,这种勇气和毅力确实不是每个团队都有的,它也是一个少数派顶着很多压力走过来的过程。

邵韦平 还好时代提供了这个机会,让我们在一个不是很成熟的设计环境中,完成了一件高难度的建筑作品。

实践

Practice

数字化

凤凰中心 [4.11]

鄂尔多斯满世广场 [4.12]

妫河建筑创意产业园规划
国际竞赛方案 [4.13]

妫河建筑创意园接待中心 [4.14]

珠海市博物馆和城市规划展览馆 [4.15]

中国尊 [4.16]

银河SOHO [4.17]

序列 [4.18]

Digitalization

Phoenix Center [4.11]
Erdos Manshi Square [4.12]
Beijing Gui River Architecture Innovation Park [4.13]
Show Room of Beijing Gui River Architecture Innovation Park [4.14]
Zhuhai City Museum and the Urban Planning Exhibition Hall [4.15]
China Zun [4.16]
Galaxy SOHO [4.17]
Sequence [4.18]

数字化：技术自由和思想自由 / 李淦

UFo对数字化技术的使用始于2006年开始的凤凰中心项目，设计的生成方式和由此产生的非线性形体迫使我们寻求不同以往的设计方法来实现这一项目。六年过去了，凤凰中心伫立于此，已从最初的一个圆润外形，发展为当下中国复杂形体建筑设计的典型案例。

凤凰中心以数字技术为平台，探索出了一套针对复杂形体建筑设计的整体控制方法与技术策略。这套控制方法与技术策略的核心是在设计、制造、建造的全过程中，运用数字技术手段，对与建筑相关的各种因素进行全面精确的整合与控制；通过建立与工程进度同步的可调控的数据信息，实现高质量的建筑语言与美学形式。我们建立了一套"三维协同"设计规则，使大家可以在同一个平台上工作，建筑师、结构及机电工程师可以基于同一个全信息建筑模型，完成设计成果的交流与传递。这种方式使凤凰中心真正实现了虚拟建造，使工程后续的生产、建造与运行工序大大受益，甚至带动了整个建筑产业链的升级和发展。

在其后的项目中，UFo广泛地采用了数字化的工作方式。非线性建筑的实现，其几何表达是一个关键问题。在银河SOHO中心项目中，为实现与传统施工方式的衔接，分为四个系统的几何定位图纸占到了建筑专业全部施工图纸的1/3，这些图纸准确而高效的生成即得益于一些数字化手段。在中国尊项目中，三维协同规则被优化成基于"建筑系统"划分构建的三维协同模式，从而有效建立起一个划分和整合工作边界的完整依据。它不仅可以使设计人员的工作成果被高效的拆分和组装，还可以保障所有需要严格控制的建筑系统均达到足够的设计深度。

Digitization: Freedom of Technology and Ideas/ Li Gan

UFo has applied digital tools from 2006 when launching a project, Phoenix Center. Owing to the design generation methods and the consequential nonlinear forms, UFo had to explore different design methods for this project. Six years passed. The Phoenix Center is now standing there. It had been developed from a simple geometry into a typical mode of complex-form architectural design in China nowadays.

Taking on Phoenix Center as a platform, UFo worked out a set of complete control methods and technical strategies for the complex-form architectural design. This setting is focused on the application of digital tools during the course of design, manufacture and construction, with an aim to carry out the complete accurate integration and control to the building related elements. Through the establishment of adjustable data in parallel with the schedule, UFo had achieved a high-quality architectural language with aesthetic forms. In addition, UFo had developed a set of "3D associate design" rules, under which all the staff were able to collaborate with each other on the working platform. That is, architects and structural engineers, as well as electrical and mechanical engineers, might exchange and transfer their design achievements on the basis of the Building Information Modeling (BIM). This made the project of Phoenix Center feature a virtual construction method, from which the subsequent production, construction and operation all could benefit. This even upgrade and drive the development of the whole industrial chain.

UFo widely applied the methodology of digital design in the subsequent works. The geometric expression is a key point to design a non-linear architecture. To make a link up to the traditional construction means in the project of Galaxy SOHO Center, UFo divided the geometrical positioning drawings into 4 systems, which together took 1/3 of all the architectural drawings in construction load. Those drawings were accurately and efficiently generated by digital tools. For the "Chinese Wine Vessel" project at CBDZ15, the "3D associate design" was optimized as a three-dimension collaborative rule that was built on the basis of the "Building System" classification, in order to efficiently form a complete basis that classifies and integrates the working limits. This not only made the working results from each designers become efficiently segmented and reassembled, but also guaranteed the strict controls of design professions required by all the construction systems.

数字技术为复杂建筑的设计提供了一个有别于传统的、更加有效和精确化的控制平台，为建筑创意的实现提供了新的技术自由。但数字技术本身绝不仅仅是静态、抽象的技术实现工具，它更大的价值和潜力是为设计的思维过程提供更多可能性，使建筑师对建筑设计的理解力和控制力进入崭新的阶段，从而使我们获得更多的思想自由。

凤凰中心的形式即产生于数字化思维：单一截面围绕复杂轨迹运动形成全新的体形。在北京妫河建筑创意产业园规划国际竞赛方案设计过程中，我们采用了一种自下而上的生形方法：利用数字化技术对场地进行分析计算，产生了随机空间、隐秘空间、非线性空间、网络化空间等创意空间，这些空间独有的不确定性最大限度地利用了土地资源，激发出人们的探求欲和想象力。在珠海市博物馆和城市规划展览馆的设计中，数字化手段被用来对不同楼层进行三维向度的连接，从而产生出多种复杂的参观流线和空间体验。

数字化技术提高了我们的控制力，拓展了我们的感知力。它所带来的技术自由和思想自由使我们不会局限于某种特定风格，从而获得更多的实践自由。

Digital tools provide a nontraditional but more efficiently and accurately designed control platform for the design of complex forms by generating ideas with technical freedom . Moreover, they provide an opportunity for architects promoting comprehension and control on design during the process of having an idea with more possible thinking freedom; and instead of static and abstract tools for implementation, they are methods of presenting greater values and potentials.

The form of Phoenix Center also originates from the digital design methodology: A single section moves along a complex trajectory to form a new geometry. For the design of the international competition proposal for Beijing Gui River Architectural Innovation Park Planning, a bottom-up forming mode system was adopted: Creative spaces like random spaces, hidden spaces, nonlinear spaces, network spaces and so on were given by means of analysis and calculation of the site via digital tools. The spaces, with the capacity to utilize the land resource, arouse exploration and imagination. The digitalization methodology was applied to make three-dimension connections between the floors of Zhuhai City Museum and Zhuhai Urban Planning Exhibition Hall, bringing about a variety of complex flow lines and spaces for visitors.

The digital tools improve the control and widen the perceptivity in UFo. It creates new design possibilities to challenge old design methods with technological and thinking freedoms.

凤凰中心

Phoenix Center

4.11

用地面积:
18 821平方米
建筑面积:
72 478平方米
容积率:
2.03
建筑高度:
52米
设计时间:
2007—2013
建造时间:
2009—2013

凤凰中心是一个集电视节目制作、办公、商业等多种功能为一体的综合建筑。通常媒体建筑都有演播厅等水平布局的大空间,也有竖向布局的办公楼标准层,这两者很难取得统一的效果。在本项目中,我们借助莫比乌斯环的图解,将高层办公区和媒体演播室融合起来,在满足全方位提供节目制作场地及其他配套服务设施的同时,形成一个完整的空间和体量,独特的建筑形态与朝阳公园自然景观有机结合为一体。凤凰中心的另一个主要特点是全方位的开放性,人们可以在其中体验媒体文化的魅力。(下图: BIM模型)

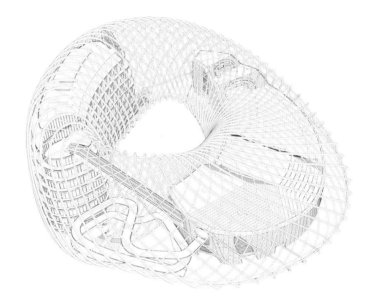

Site Area:
18,821m²
GFA:
72,478m²
FAR:
2.03
Building Height:
52m
Design Time:
2007-2013
Construction Time:
2009-2013

The Phoenix Center building is a mix-used development with various functions including TV-programming, office and retail. Usually, a media architecture incorporates a horizontal spreading space volume and vertical levels, for example, a big studio space and typical office levels, but it is difficult for the two plans to achieve a coordinative effect. In this project, we used a Mobius Band to link high-level offices with the media studio, in order to create a coherent space volume while providing the comprehensive programming arena with the surplus service facilities. While nicely fit into the urban fabric of Chaoyang Park, the building's omnibearing openness would enable visitors to enjoy its charming media culture.
(Above: BIM model)

凤凰中心的创意来源于莫比乌斯环,其体形既连续又变化,复杂的三维空间把城市界面和公园界面统一在一起。同时,三维曲面外壳把南侧高层办公楼和北侧演播楼连成整体,并在东西两侧形成两个空中大堂。立面采用金属结构作为遮阳板和玻璃幕墙相结合的方式,起到节能的作用,同时反映了建筑造型的灵动感。通过数字化技术将科技和艺术表达完美结合。

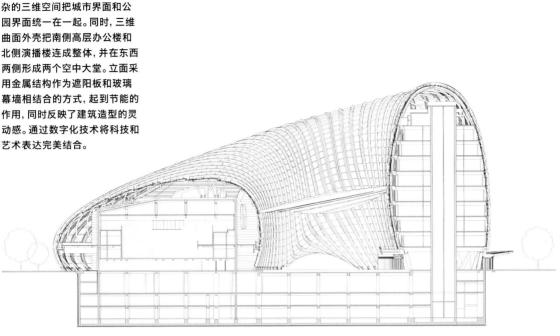

The creative idea of Phoenix Center originates from Mobius Strip. Its shape is both continuous and changeable. The sophisticated three-dimensional space unifies urban interface and park interface. Meanwhile, its three-dimensional surface crust links into a whole entity the high-rise office buildings at the south and the studios at the north, and forms two sky lobbies both at the east and at the west. The elevation adopts a metallic structure to combine the sun shields and the glass curtain walls in order to save energy and reflect the spirit of the building structure at the same time. Herein, science and artistic expression are perfectly combined through digital technology.

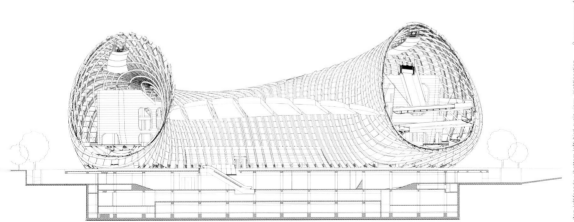

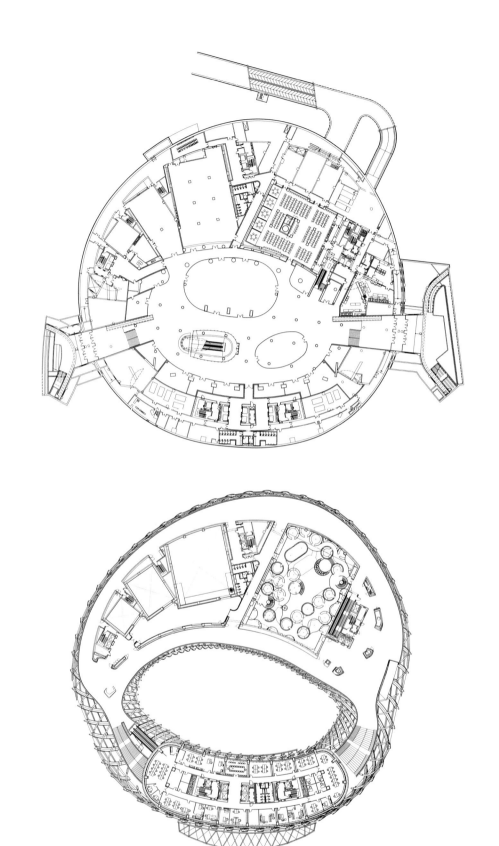

(top-left)B1 plan; (bottom-left)2F plan; (right)bidirectional super-positioning grid structure system   0———10m   (左页上图) 地下一层平面图; (左页下图) 二层平面图; (右页图) 双向叠合网格结构体系

建筑布局朝向公园开放,人们可以穿过基地进入公园,建筑中开放的参与互动活动也与公园融为一体。建筑采用柔和的流线形造型,符合空气动力学原理,各个方面均保持了光滑连续的界面,减小与环境的冲突,避免街道风的形成,具有生态建筑的特征。

The project's layout is open to the park area, whose visitors can make access through the base building to the site. Open activities are accessible to those visitors within a same interactive atmosphere from the park. An elegant curving form, which complies with aerodynamic principles is adopted by architects. While maintaining smoothness and continuous interfaces, the building reduces its confliction within the context and avoid wind qusts as an ecological design strategy.

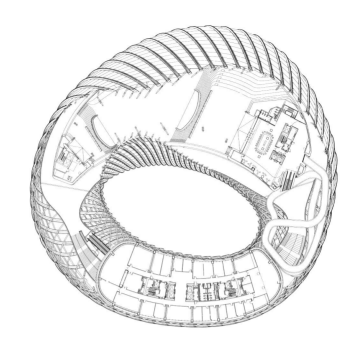

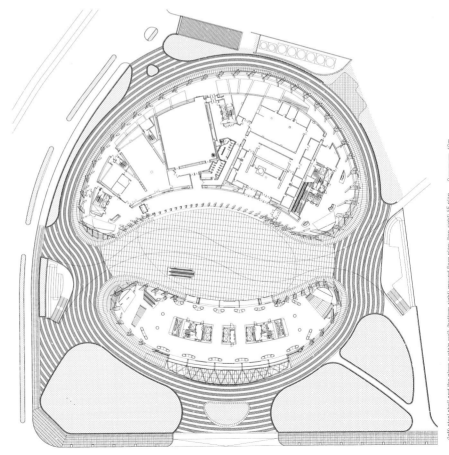

(left) steel shell and the glass curtain wall; (bottom-right) ground floor plan; (top-right) 5F plan

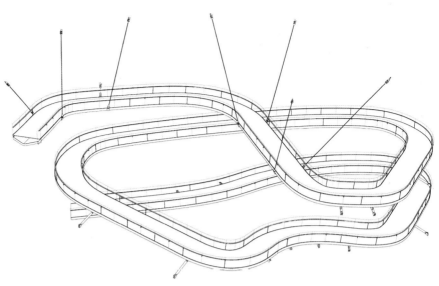

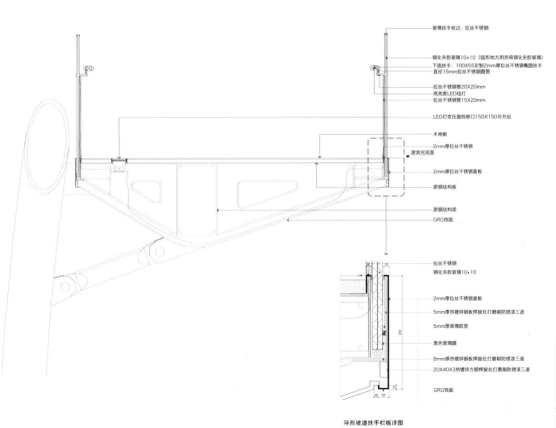

环形坡道扶手栏板详图

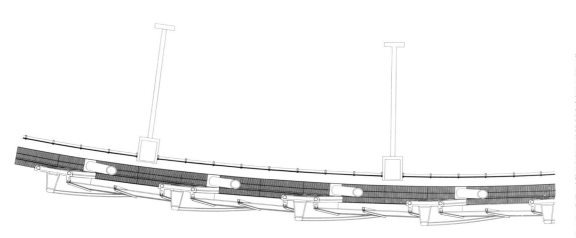

(top-left) MCM 2015 Fashion show,Phoenix center; (bottom-left) east atrium; (right) detail of glass curtain wall

(左页上图) MCM2015春夏系列全球首秀在凤凰中心举行; (左页下图) 东中庭; (右页图) 外幕墙节点

鄂尔多斯满世广场

Erdos Manshi Square

4.12

用地面积:
7 830平方米
建筑面积:
146 279平方米
容积率:
4.27
建筑高度:
111.5米
设计时间:
2009—2011
建造时间:
2011—

满世商务广场位于鄂尔多斯市东胜铁西区核心地段,场地约160平方米。北侧较高,南侧较低,最大高差约为四米。东侧为东环路,西侧为城市集中绿化带,南侧为规划路,北侧是科技街。该项目功能配置包括商业写字楼、超市以及地下车库三种主要类型。整体空间布局分为南北两个街区,中间形成商业街。两座塔楼均朝向城市转角微微倾斜,有更大的展示面,同时避免朝北的立面全天都处在阴影之中,形成丰富的光影效果。高层建筑的竖向百叶组成抽象的世界地图图案,呼应本项目案名。

(下图: 世界地图—立面分析图)

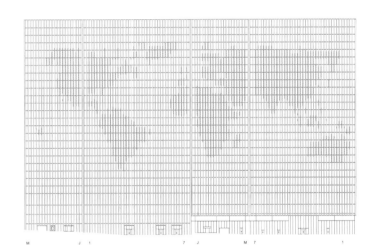

Site Area:
7,830m²
GFA:
146,279m²
FAR:
4.27
Building Height:
111.5m
Design Time:
2009-2011
Construction Time:
2011-

Located in the downtown of Dongshengtiexi District, Erdos City, the Manshi Business Square occupies an area of 160 square meters, whose north side is higher than the south within the maximum elevation difference of 4 meters or so. For this square, East Ring Road, a key green zone, Guihua Road and Science Street together form borders from the east to the north in clockwise. The functional configuration of this project includes such three main types as business office buildings, supermarkets and underground garages. The whole spatial layout is divided into two blocks, with the middle formed as a commercial street. Both of the two tower buildings slightly face the city corner so as to enjoy more display surface and at the same time avoid that the elevation facing the north would remain in shadow for the whole day, which of course shows ample sunlight and shading effects. The vertical louvers of a high-rise building form an abstract pattern of world map echoing the project title (Manshi means "the whole world" in Chinese). (Above: The world map-Elevation analysis)

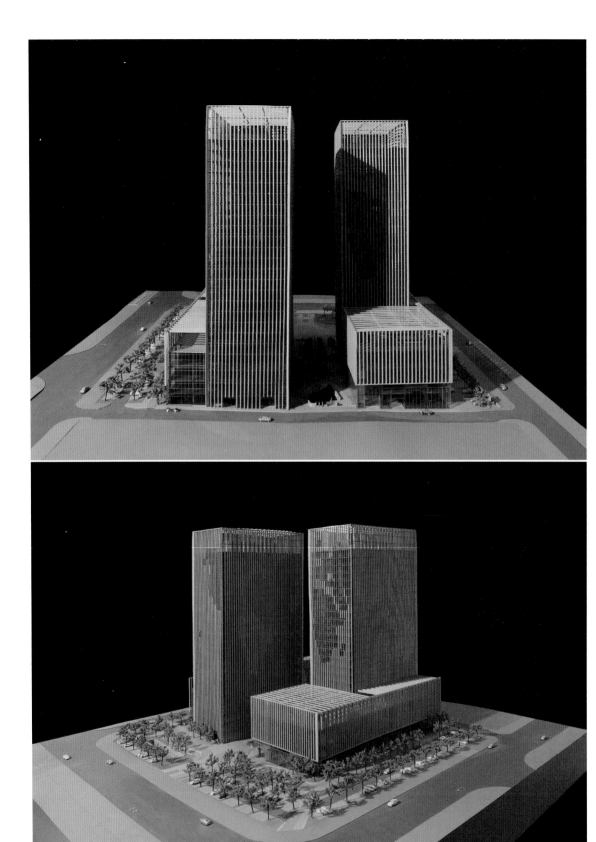

竖向百叶通过数字化手段组合成抽象的世界地图图案，在单一竖向构图的条件下实现建筑外观的丰富性，同时呼应本项目案名。

景观设计以自由流畅的曲线为母题，与方折冷峻的建筑形象互相衬托，相得益彰。

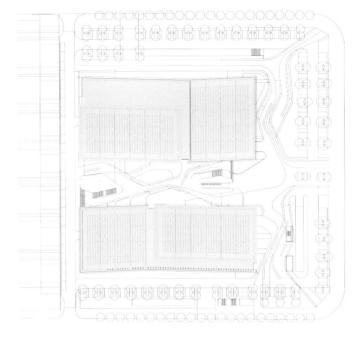

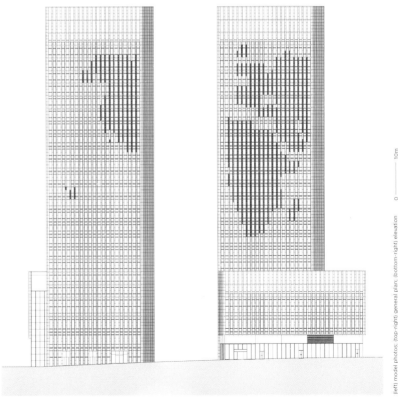

By digital design methods, vertical louvers are designed to form a pattern of an abstract world map, which enriches the facade using a single design elements.

The landscape design uses a free and smooth curve as its motif which is mutually set off by the square and forbidding architectural image, which brings out the best in each other.

1	铝横框
2	2mm 铝单板
3	保温岩棉
4	连接支座组件 热侵镀锌
5	铝竖框
6	铝合金副框
7	一体化透镜
8	铝合金灯槽
9	石材龙骨角 钢架组件
10	石材
11	灰色石材密封胶
12	铝合金托板
13	镀锌钢板
14	中空玻璃

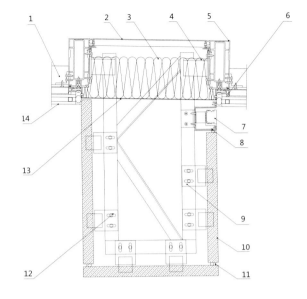

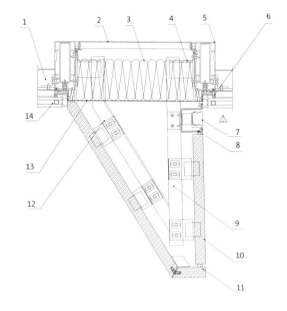

1	Aluminum frames
2	2mm Aluminum sheet
3	rock wool insulation
4	Hot dip galvanizing bearing component
5	Aluminum mullions
6	Aluminum alloy sub-frame
7	Integrated lens
8	Aluminum alloy troffers
9	steel frame for stone finish
10	Stone finish
11	gray Stone sealant
12	Aluminum alloy supporting plates
13	Galvanized steel sheets
14	Insulating glass

经过大量的细部研究比较,我们选定了两种典型的外墙立挺节点,从各方位观看建筑都能获得不同的微妙差异。这一处理可以保证每个房间采光条件的均质化和外窗的标准化,同时夜景照明等相关的细节需求也被整合到构造中。

After large amount of research and contrast and comparison in detail, we have decided on two typical curtain wall details so that subtle differences can be seen no matter from which direction it is viewed. This type of handling can make sure the homogenization of lighting conditions and the standardization of exterior windows at each room. At the same time, nightscape lighting and other such detailed requirements are also integrated into the structure.

(左页图)两种不同幕墙竖挺节点;(右页图)施工现场照片

(left) curtain wall mullions are two different nodes; (right) construction site photo

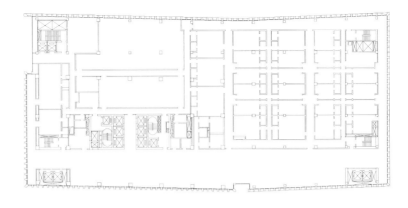
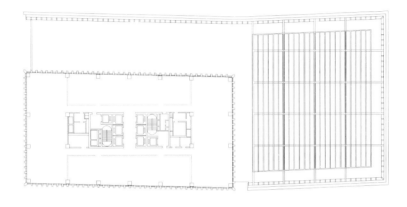
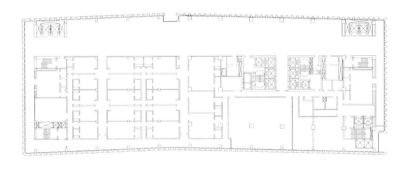
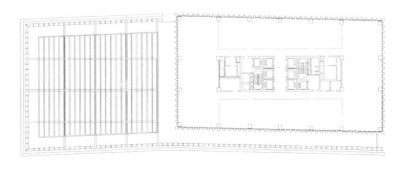

(left) renderings; (right, top to bottom) 3F and 7F plan of A zone, 3F and 7F plan of B zone

妫河建筑创意产业园规划
国际竞赛方案

Beijing Gui River Architecture Innovation Park

4.13

用地面积:
210 150平方米
建筑面积:
145 000平方米
容积率:
0.8
建筑高度:
54.2米
设计时间:
2009
建造时间:
2010—2012

北京妫河建筑创意产业园位于延庆,是以建筑创意为主产业的高质量创意产业园区,将为泛建筑创意领域的创意型企业(或组织)提供创作、培训、科研、成果展示、文化交流的场所。方案避免了开发区"先破坏、后修复"的建设模式,将基地打造成一个集湿地、林地、坡地、台地为一体的郊野公园,在"智慧藏于自然"的东方哲学观念中,成为滋养创意活动的天然温床。综合利用风、水、太阳、地热等,低碳的活动模式将重新定位人与自然的关系。这个方案代表北京市建筑设计研究院参加了该项目的国际设计竞赛。(下图: 活动区域分析图)

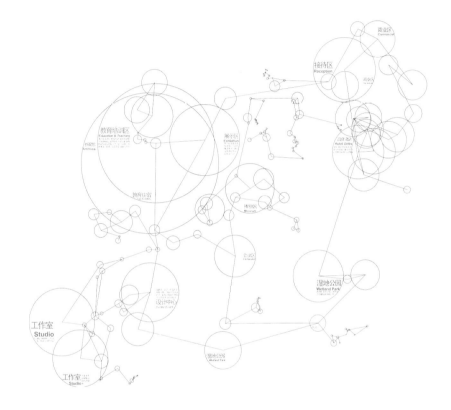

Site Area:
210,150m²
GFA:
145,000m²
FAR:
0.8
Building Height:
54.2m
Design Time:
2009
Construction Time:
2010-2012

Beijing Gui River Architectural Innovation Park, Yanqing, Beijing, thrives to become a quality building-idea base, in order to provide a variety of creative construction enterprises or organizations with the sites of composition, training, research, presentation and culture exchange. Depending on a proposal without the building idea of "destruction first, restoration later", the base is constructed as a surburb park integrated with wetlands, woodlands, sloping fields and terraces. With a tenet "wisdome is hidden in nature", which is an oriental phylosophy, the base becomes raising the creative activities on the comprehensive utilization of wind, water, solar, and geothermal energy. Such the low-carbon activities will re-position the relationship between human and nature. This was a competition proposal on behalf of BIAD. (Above: Activity area analysis)

方案从四个方面最大限度地体现了创意产业园的聚集发展特征。开发模式：能满足各种功能需要，建筑体量之间没有过多的制约关系，位置、大小、材料、高度等各个方面都有很大的发挥空间。开发周期：方案在实施的任何阶段都有完整的形象，随着时间的发展能够自行调节进度，在规划最终完成后仍有很大的自我发展空间。空间形态：建筑、外部空间、自然环境一气呵成，空间层次非常丰富，具有小镇的宜人尺度，便于开展各种活动，具有旅游价值。开发成本：采用小尺度建筑聚合成整体的布局，因此避免过大的集中开发，使开发成本易于控制。

The proposal presents the Park's cluster feature in four aspects at best. Development Schema: Be able to meet various functional needs, without too much restrictive correlation between building masses. Be able to have much development space in such aspects as location, size, material, height and so on. Development period: The plan should have a complete form at various stages of implementation. As time goes, it can adjust progress by itself so that it remains to have much space for self-development when the planning is finally completed. Spatial Form: There is much coherence among buildings, exterior space and natural environment. It is fraught with enough spatial levels, with as many pleasant scales as a village has, so that it is easy to carry out various kinds of activities, which shows its tourism value. Development cost: It adopts such a layout as small-scale buildings integrate into a whole entity so as to avoid oversized centralized development and keep the development cost easy to control.

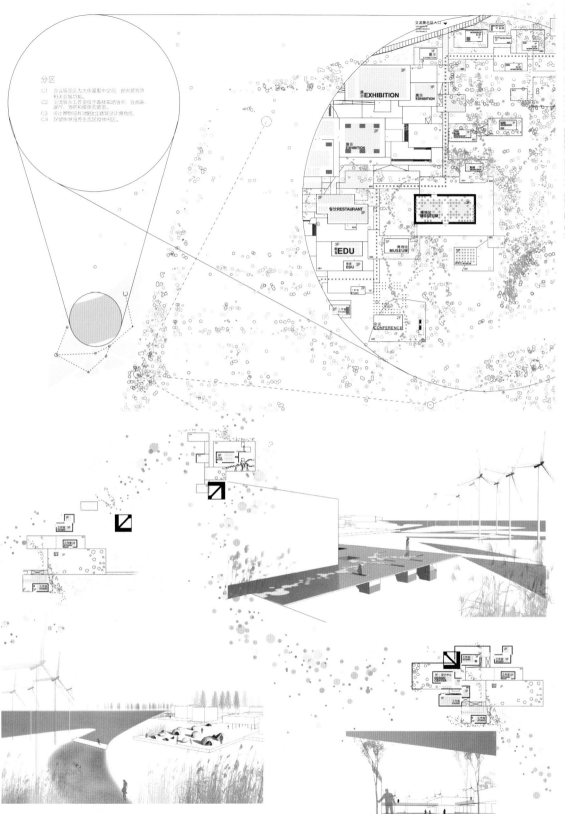

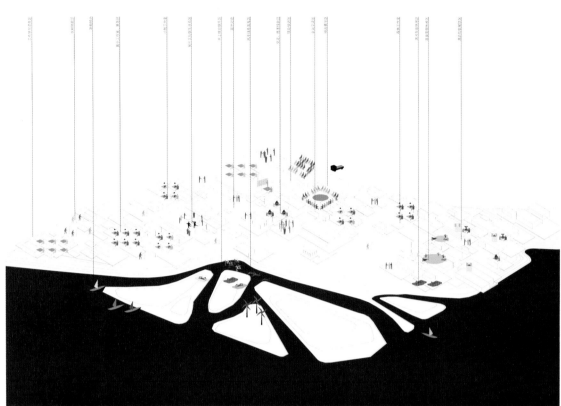
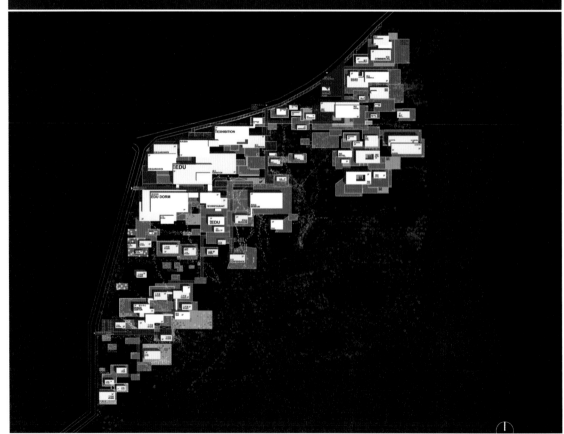

利用数字化技术对场地进行分析计算,最大限度的利用了土地资源,产生的随机空间、隐秘空间、非线性空间、网络化空间等创意空间具有不确定性,将激发人们的探求欲和想象力。

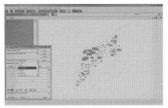
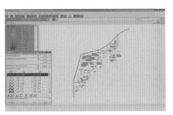
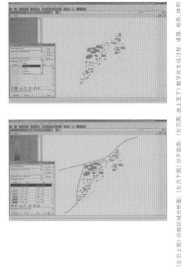
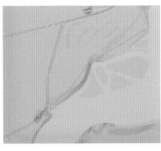
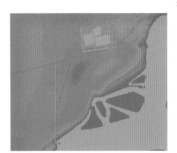
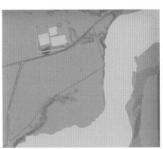
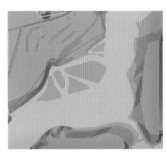

The proposal applies digital tools for using the land resources by the maximized on-site analysis and calculation. This generates the uncertainty of creative spaces, such as a random space, a hidden space, a nonlinear space and a network space, all of which will inspire people to exert themselves with desire and imagination for knowledge.

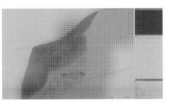
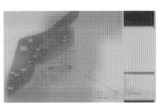

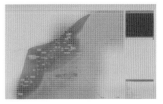

妫河建筑创意园接待中心

Show Room of Beijing Gui River Architecture Innovation Park

4.14

用地面积:
1 361.6平方米
建筑面积:
3 595平方米
容积率:
0.81
建筑高度:
9.8米
设计时间:
2009—2011
建造时间:
2013

妫河建筑创意园位于北京延庆县妫河北岸延庆镇西屯村妫河北岸,占地面积21公顷,总建筑面积14.5万平方米,是一个以建筑创意产业为主,集设计研究、教育培训与展示交流的创意平台。它的规划建设意在提高延庆的环境吸引力。我们将基地打造成一个集合了湿地、林地、坡地、台地的郊野公园,使未来的创意产业园成为一个开放的服务平台,兼容各种个性化团队的需求,也避免发展单元之间的相互制约。接待中心位于园区01-06-22地块,是创意区的首个开发项目,主要功能为接待、展示、办公、餐饮和住宿。工程用地面积2 957平方米,建筑面积3 595平方米。建筑地下一层,地上二层。整个造型在通透的首层上方用挑台托起二层错落的客房,形成"漂浮的村落"的意向。这一形态呼应了水畔建筑的小尺度特征,也延续了规划设计中灵活变化的"簇群式"功能组合和随机空间、隐秘空间、非线性空间、网络化空间等创意空间独有的不确定性。UFo负责此项目从规划和城市设计到建筑单体和室内精装的全程设计。(下图: 项目位置)

Site Area:
1,361.6m²
GFA:
3,595m²
FAR:
0.81
Building Height:
9.8m
Design Time:
2009-2011
Construction Time:
2013

The Architectural Innovation Park is situated on the north bank of Gui River, Xitun Village, Yanqing Town, Yanqing, Beijing, within an area of 21 hectares and a total GFA of 145,000 square meters. This park is operated as a platform engaged in the creative buildings oriented R&D, education, presentation and exchange. Its planned construction intends to promote Yanqing's environment attraction. The proposal maker wishes to build the base into a suburb park in combination with wetlands, woodlands, sloping fields and terraces. In the future, the creative park will become an open service platform incorporating the requirements of various personalized teams, so as to avoid the mutual restrictions of development units. Information Center is at No. 01-06-22 in the park. As one of the park's first batch of development projects, this center provides comprehensive services in reception, presentation, office, food and accommodation. The work occupies an area of 2,957 square meters and sees a GFA of 3,595 ones, covering an apartment level and two ground floors. In the fabric, a cantileverred terrace over the first floor supports the overlapped guest rooms on the second floor, with an aim to build the "floating villige" for making response to the small-scale characteristic of waterfront building. In addition, the flexible "cluster" of features in a plan is continued along with the unique uncertainty of creative spaces, such as the random, secracy, nonlinear and network ones. UFo is undertaking an overall course on the planning, the urban design, the individual unit and the indoor decoration. (Above: Project Location)

设计概念来自于功能和形式的矛盾：如何在满足展示接待对大空间的功能需求的同时，保留形式上的小尺度特征？我们将原本在地面上的"村宅"提升到+5m标高，其下就获得了开敞、完整的展示与接待空间。日常形态的非正常位置带来了建筑的异常感，+5m也成为建筑上下两部分不同材质和处理手法的分界线：下部采用玻璃和金属等清透材料；上部反而采用陶土砖这一厚重材料，以强化"漂浮的村落"的别样体验。电影《武侠》的这张剧照很好地诠释了这一设计概念。

The design concept comes from the contradiction of function and form: how to show the functional requirement that guest receiving has on large space while keep small-scale characteristics that form requires? The conclusion is to elevate the "village residence" originally on the ground into the standard height of +5m so that the space underneath would enjoy open, complete display and receiving space. Off-normal position of daily form brings about a sense of abnormality for the building itself, and +5m thus becomes the dividing line between different materials and handlings for the upper and the lower portions of the building. That is, the lower portion adopts glass, metal and other such clear materials, while the upper portion uses such solid material as terracotta bricks so as to emphasize the different experience of "floating village." This picture of the movie "Martial Arts" perfectly interprets such a design concept.

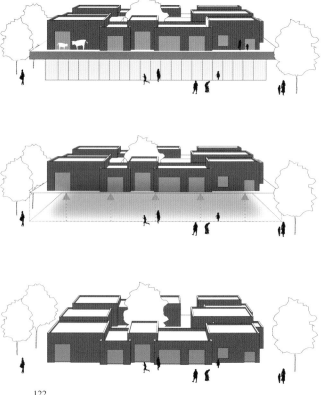

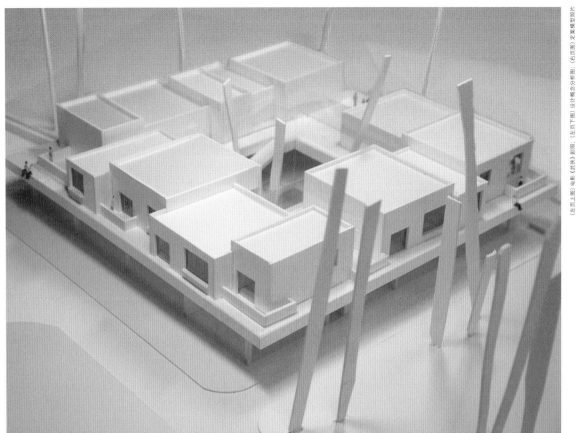

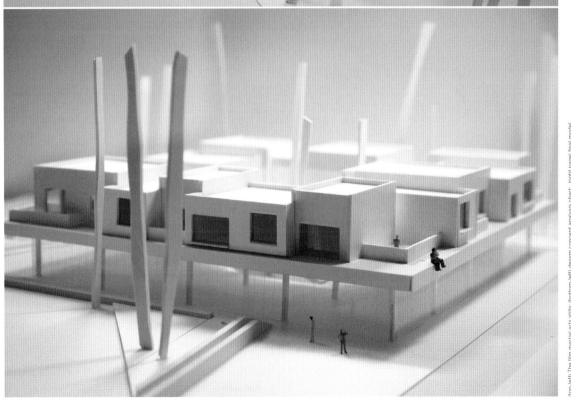

(top-left) The film *martial arts* stills; (bottom-left) design concept analysis chart; (right page) final model

(左页上图) 电影《武映》剧照；(左页下图) 设计概念分析图；(右页图) 定案模型照片

数字化手段产生了海量的体块选择，最终确定的体块满足了以下标准：客房单元的标准化、上下结构搭接的简单化和形体的丰富性。

The digital means result in volumes of choices on body masses. The body masses that are finally decided on meet the following standard: standardization of the guest room unit, simplification of the joining of the upper and the lower structures, and richness in form.

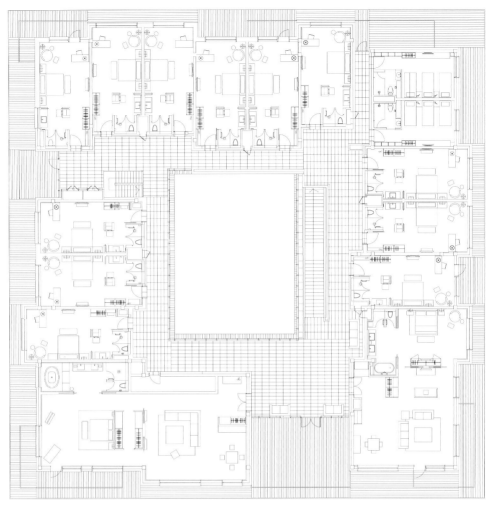

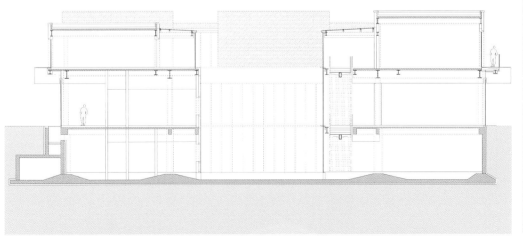

(left) second floor guest room layout comparison; (bottom-left) site plan; (top-right) 2F plan; (bottom-right) cross-sectional view

(左页上图)二层客房体块比较;(左页下图)总平面图;(右页上图)二层平面图;(右页下图)剖面图

二层客房外墙使用了不同颜色的陶土砖以体现这一项目应有的轻松感。按照客房单元体的高度（3.80米和4.80米两种类型）和宽度（单、双开间两种类型）分为单低、单高、双低、双高四种体块类型，分别对应四种砖色。明晰的逻辑取代了人为的美学判断。

The envelope of guestroom units on the second floor applies terracotta bricks in different colors to make the easy feeling of project. Depending on their two heights (3.80m and 4.80m) and two widths (single and double), those units are in 4 mass types, the single low, the single high, the double low and the double high, which are aligned with four brick colors, respectively. Therefore, a logic of perspicuity replaces the human aesthetic judgment.

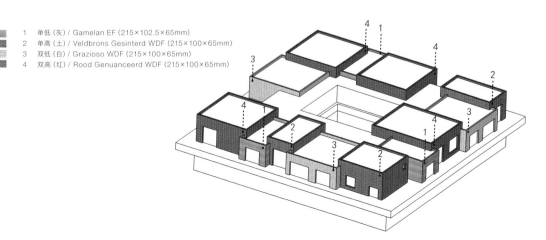

	1	单低（灰）/ Gamelan EF (215×102.5×65mm)
	2	单高（土）/ Veldbrons Gesinterd WDF (215×100×65mm)
	3	双低（白）/ Grazioso WDF (215×100×65mm)
	4	双高（红）/ Rood Genuanceerd WDF (215×100×65mm)

(左页下图) 陶土转材质选择；(右页下图) 外墙材质分析图；(上图) 外观
(bottom-left) clay transfer material selection; (bottom-right) exterior wall material analysis; (top) appearance

127

首层吊顶研究过程：根据人的活动区域和路线，设置顶部照明强弱范围，采用八种规格的采光吊顶单元，实现吊顶图案和相应光环境的非线性变化。最终的吊顶设计，看似错落斑驳的风格同样也出自严密的几何逻辑。

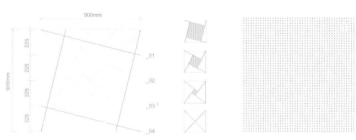

Research process of suspended ceiling at the ground floor: according to activity areas and routes, scope of strength in overhead lighting is set up. It adopts suspended ceiling lighting unit of 8 specifications in order to realize nonlinear variation between the graphic patterns at the suspended ceiling and the corresponding luminous environment. The seemingly random and mottled style of the final design of the suspended ceiling is also based on strict geometric logic.

1 开敞办公区 / Open office area
2 主馆办公室 / Main hall office
3 员工用餐区 / The staff dining area
4 会客区 / The reception area

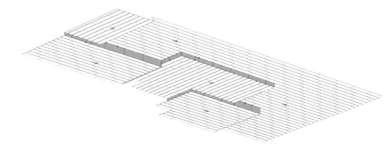

珠海市博物馆和城市规划展览馆

Zhuhai City Museum and the Urban Planning Exhibition Hall

4.15

用地面积:
50 335平方米
建筑面积:
48 400平方米
容积率:
0.72
建筑高度:
54.2米
设计时间:
2009

珠海市博物馆和城市规划展览馆是珠海市地方性综合馆,两馆的展览空间为错层布置,既有互相渗透的开放性,又能保持各自的独立性。垂直交通核心区利用错层的叠合空间,充分利用了空间的高度,布局更加紧凑,使用和管理都很高效。博物馆和规划馆设计亦分亦合。两馆的功能适当共享而不是完全分离。在排布展区时,不仅考虑到每个馆自身参观流线的组织,也考虑到两个展馆之间的参观流线衔接,从而把历史、现在、未来纳入到一个统一的序列中,将大循环的时空关系依次展现在参观者面前。博物馆从上到下按时代顺序排列,依次是历史展区、专题展区、特殊展区、临时展区、改革开放展区;规划馆从下到上按照从总体到局部顺序排列,依次是宣传教育区、城市历程、总体规划、基础设施、大规模展区、重点地区和各区规划展示。(下图: 分区示意图)

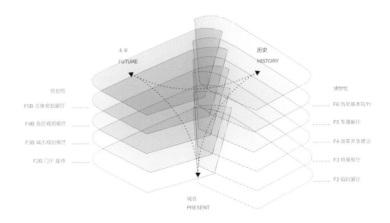

Site Area:
50,335m²
GFA:
48,400m²
FAR:
0.72
Building Height:
54.2m
Design Time:
2009

Zhuhai Municipal Museum and Urban Planning Exhibition Center are Zhuhai's local general exhibition halls. The exhibition space of both places is arranged at split level, open to mutual penetration while able to keep their own independence. The core zone of vertical transportation takes the advantage of spatial overlay at split level, making full use of the height of the space to better tighten the layout, which results in efficient utilization and management. The designs of the Museum and the Exhibition Center are both separate and together. The functions of both halls are properly shared, instead of totally separated. When arranging and laying out the exhibition areas, not only the organization of line of visitors at each hall has been taken into consideration, but also the connection of the line of visitors between both halls has been thought of, as well. As a result, the ideas of the history, the present and the future can be incorporated into a unified sequence, which successively shows visitors the time-space relationship of systemic circulation. From top to bottom, the Museum is arranged in such time order as the History Wing, Theme Display, Special Exhibition Area, Temporary Exhibition Area, and Reform and Opening-up Exhibition Area. From top to bottom, the Urban Planning Exhibition Center is arranged in such an order as from the overall to the localized, which is shown as Publicity and Education, Urban Advancement, Overall Planning, Infrastructure, Large-scale Exhibition Area, Key Region and Regional Planning.

(Above: Schematic partition)

博物馆参观流线
如果只参观博物馆，则乘坐景观电梯到达顶层，顺坡道下至各层参观。参观顺序是从历史到现在。

规划馆参观流线
如果只参观规划馆，则由首层上至各层，从顶层乘景观梯下。参观顺序是由总体到局部重点地区。

博物馆—规划馆参观流线
如果要参观两个馆，第一个选择是由博物馆至规划馆：乘坐景观电梯到达顶层，顺坡道下至博物馆各层参观；到达二层门厅后再由规划馆首层上至各层，从顶层乘景观梯下。参观顺序是从历史到现在再到未来。

规划馆—博物馆参观流线
第二个选择是由规划馆至博物馆：由规划馆首层上至各层，从顶层顺坡道下至博物馆各层参观。参观顺序是从现在到未来再追溯历史。

博物馆参观流线 /
Visiting route of the Museum

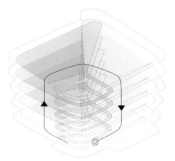

规划馆参观流线 /
Visiting route of the Urban Planning Exhibition Hall

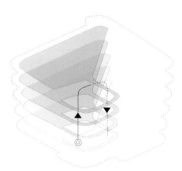

博物馆—规划馆参观流线 /
Museum-Museum visits streamline planning

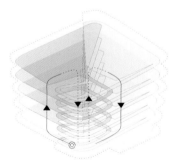

规划馆—博物馆参观流线 /
Planning Museum-Museum flow lines

Route in Zhuhai City Museum
If only for the museum, the visiters will take a landscape elevator to the top floor and then descend along a ramp to the lowers, taking a turn to exhibition rooms from the historic to the present.

Visiting route of the Urban Planning Exhibition Hall
If only for the hall, the visiters can go upstairs from the first floor to the above and then descend from the top floor by the elevator, taking a turn from the overall to the specific locals.

Visiting route of both buildings
If for both, the first visiting option is to begin from the museum: Take a landscape elevator to the top floor and then descend along a ramp to each a lower level in the museum; by the second floor's lobby, access the hall from the first floor and at last descend from the top by the elevator, taking a turn from the historic to the present and then the future.

Visiting route of Planning Hall-Museum
The second option is to begin from the hall: Ascend from the first floor to the other levels in the hall; and then descend along a ramp to all the floors in the museum, taking a turn from the present and future to the historic.

参观流线经过精心编排：博物馆顺坡道下至各层参观，以及规划馆顺坡道上至各层参观，行进方向均面朝大海。

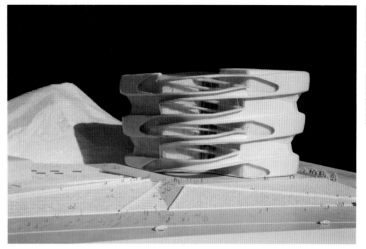

All of the visiting routes are elaborately arranged: When making access along the ramp, either descending or ascending, the forwarding directions are both in the face of sea.

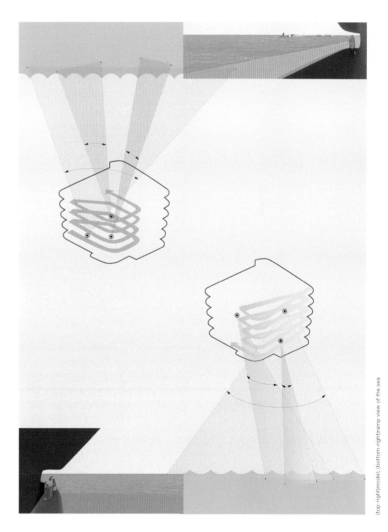

中国尊

China Zun

4.16

用地面积:
11 478平方米
建筑面积:
437 000平方米
容积率:
30.5
建筑高度:
528米
设计时间:
2011—2014
建造时间:
2013—

北京市朝阳区CBD核心区Z15地块超高层项目,占据北京CBD中心和中央公园的关键位置,基地为136米×84米的矩形,其中东、西、北三面分别紧临金和东路、金和路及光华路,南侧红线与规划中的艺术中心项目基地衔接。该项目是一座集办公、高档会所以及公众观光等多功能于一体的超高层城市综合体,建成后将是北京重要的地标性塔楼,为北京的城市天际线增添新的巅峰。塔楼造型灵感取自中国传统礼器"尊",经过抽象处理和比例优化,既保持了"尊"形独特的弧形轮廓效果,又形成比例上的优雅和秀美。作为北京最高的多功能超高层,Z15塔楼采用领先高效的安全结构,使用高新技术进行分析,设置多重设防抗侧力体系,提供多重竖向传力体系,增加坚固性,同时保障使用者的舒适度。UFo在设计中担当项目总负责及结构外框筒与玻璃幕墙系统几何形态的设计控制与协调、地下室空间(B2-B7层)以及总平面设计。(下图:总平面图)

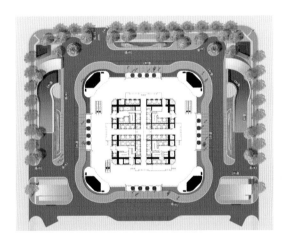

Site Area:
11,478m²
GFA:
437,000m²
FAR:
30.5
Building Height:
528m
Design Time:
2011-2014
Construction Time:
2013-

Within the CBD in Chaoyang, Beijing, the No. Z15 project of ultra high-rises featured a key location in both the CBD and the central park, within a square base of 136m times 84m in size. Its east, west and north sides were Jinhe East Road, Jinhe Rode and Guanghua Road, and its south gets close to the base of a planned art center. The No. Z15 is also a complex of ultra high-rises in integration of the features of office, high-level club and sightseeing. It would become one of key landmark towers in Beijing's skyline. In conception, the facade of tower had simulated a Chinese traditional wine vessel - Zun in name, which had been abstractly processed and proportionally optimized, so its unique curved outline was remained and the proportional elegance and grace came into being. As one of the tallest mix-use buildings in Beijing, the Z15 tower applied an advanced and efficient safety structure with an advanced technical anaylsis. For the same time, it was equipped with multiple side-stress-proof systems and vertical-dynamic-transition systems to increase robustness while satisfying the comfort of users. During the project workout, UFo took the responsibility of designing, controlling and coordinating the structural outside-frame tube and the glaze-screen-system geometry, with providing the solutions of basement space and general plan. (Above: site plan)

该项目总体设计框架以建筑的巨框结构和十个功能分区为前提条件，考虑到规模、交通、消防、安防、气象等因素，划分五个单元模块。其中ZB区与Z0区为模块一（中枢模块）；Z1区与Z2区为模块二（功能模块）；Z3区与Z4区为模块三（功能模块）；Z5区与Z6区为模块四（功能模块）；Z7区与Z8区为模块五（功能模块）。

The overall design framework took the building's giant box structure and ten functional divisions on premise. This covers five modules in scale, transportation, fire prevention, security and climate as follows:
ZB division and Z0 division are Module I (central module);
Z1 and Z2 are Module II (function module);
Z3 and Z4 are Module III (function module);
Z5 and Z6 are Module IV (function module);
Z7 and Z8 are Module V (function module).

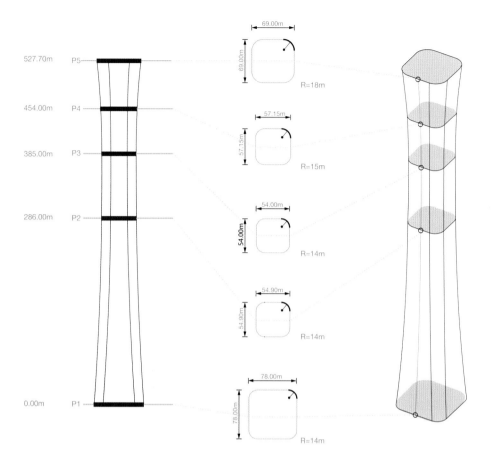

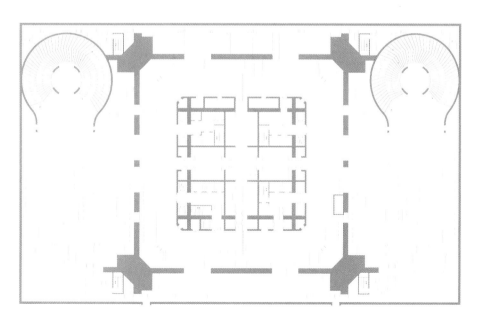

(left) BIM model; (top-right) chamfering of basic control surface; (bottom-right) B3 plan

国际化的设计团队与设计标准为该项目的高品质提供了保障。基于建筑系统划分的BIM技术的引入和PW协同工作管理平台的建立，有效提高了团队协作的工作效率，保证了设计成果的整合性和精确性。

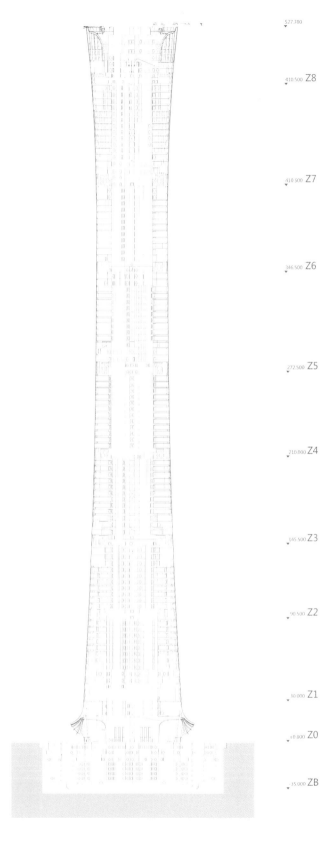

527.700
410.500 Z8
410.500 Z7
346.500 Z6
272.500 Z5
210.000 Z4
145.500 Z3
90.500 Z2
30.000 Z1
10.000 Z0
15.000 ZB

The project saw a high quality thanks to the transnational design teams and standards. The building-system-class-based BIM technology was introduced and the PW management platform for coordinative working was formed. This secured both the efficiency of teamworks and the integrity and accuracy of design results.

基础控制面
Basic control surfaces

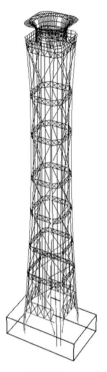

基础控制线
Basis of the control line

巨柱
Pillars

腰桁架斜撑
Waist truss bracing

重力柱内尊
The statue of gravity column

核心筒楼板
Core tube floor

二次钢结构
Secondary steel

幕墙
Curtain

银河SOHO

Galaxy SOHO

4.17

用地面积:
52 546.785平方米
建筑面积:
330 117平方米
容积率:
4.64
建筑高度:
60米
设计时间:
2008—2010
建造时间:
2010—2012

银河SOHO中心是一个集商业、办公于一体的大型综合项目,由英国扎哈·哈迪德建筑师事务所完成方案设计和外立面的技术设计;BIAD完成建筑专业平面系统初步设计和结构、机电专业初步设计以及全部施工图设计。UFo与其他专业一起从空间、专业整合、细部构造等各方面优化了原设计:采用心形核心筒形式和周边柱型式,给内部空间带来巨大改观;创造性采用多种技术措施,高效解决了复杂的消防安全问题;综合各专业复杂因素,在满足各专业要求的同时,保证建筑外观的高度整洁;对于此类高难度非线性建筑,使用数字化手段进行精确的几何控制,并且实现数字化方式高效准确出图;搭建全专业BIM模型,有效控制了施工和设计的全过程,使银河SOHO中心项目成为当代数字化技术应用的典范。本项目已获2013年英国皇家建筑师学会国际大奖。(下图: 总平面图)

Site Area:
52,546.785m²
GFA:
330,117m²
FAR:
4.64
Building Height:
60m
Design Time:
2008-2010
Construction Time:
2010-2012

Galaxy SOHO Center is a mix-use development including retails and offices. The schematic and facade design are provided by Zaha Hadid Architects. BIAD was responsible for the design development of architectural plans, structural system, and MEP design, as well as construction documents. UFo, together with other professions, optimizes the original design in such aspects as space, system integration, details and so on. That is, it adopts such new models as heart-shaped core and perilmeter colulmns in order to drastically improve interior space. UFo creatively adopts multi-tech measures, effectively solves the complicated fire and safety problems, synthesizes the complex factors of various systems, so as to ensure the cleanness and tidiness of the exterior of the building while satisfy the requirements of various systems. In light of such highly complex nonlinear buildings, UFo adopts digital measures to take accurate geometric control, and realizes efficient and accurate digital output. It builds full-scale professional BIM models, and effectively controls the whole process of both construction and design in order to set Galaxy SOHO up as an example of contemporary digital technology application. The project has won an internationalaward from the Royal Institute of British Architects (RIBA). (Above: site plan)

从建筑功能上，地下一层至三层为商业，四层至十五层为办公，地下二、三层为车库和机房。三层以下设室内中庭，三层以上为室外庭院。

项目包括四个卵形建筑，各卵形体之间在不同楼层由高低区连桥相连，从而组成一栋大底盘多塔建筑。通过单体的整合，营造出一个壮观的银河系星云状整体。

As for functions, B1 to F3 are retails, F4 to F15 are offices, B2 and B3 are parkings and mechanical rooms. Interior atriums are designed for floors below F3, and outdoor courtyards are for higher floors.

The project contains four egg-shaped buildings, which are linked by connecting bridges in high and low sections on different floors to form a multi-tower building with an enlarged base. This integration of individual units creates a spectacular galaxy.

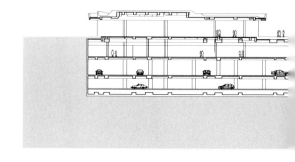

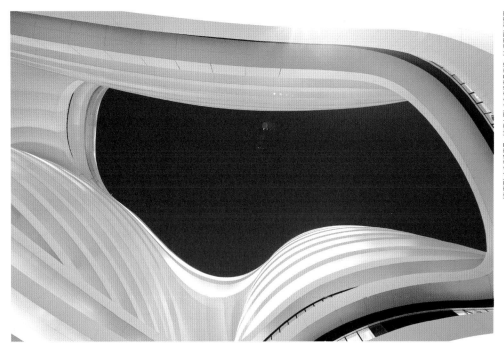

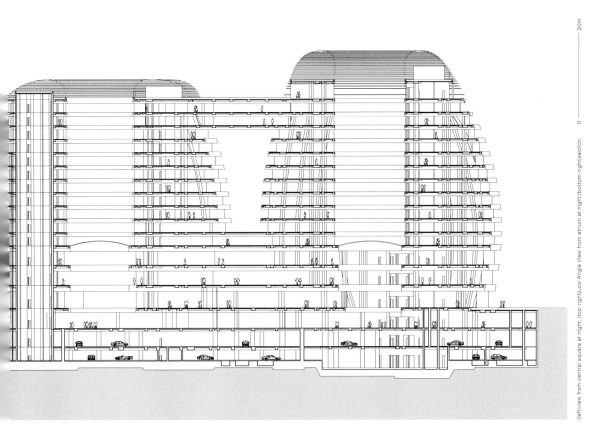

(left)view from central square at night;(top-right)Low Angle View from atrium at night;(bottom-right)section (左页图)中央广场夜景;(右页上图)中央广场夜景仰视;(右页下图)剖面图

由于项目形体的复杂性，建筑专业施工图中超过1/3的内容为几何定位图纸，分为柱及轴线、结构板边、外幕墙和内幕墙四个系列。

Owing to the compexity of the project's forms, 1/3 of the construction documents are the geometrically positioning drawings which are divided into four series, the column and axis, the structural plate edge, the external envelope and the internal envelope.

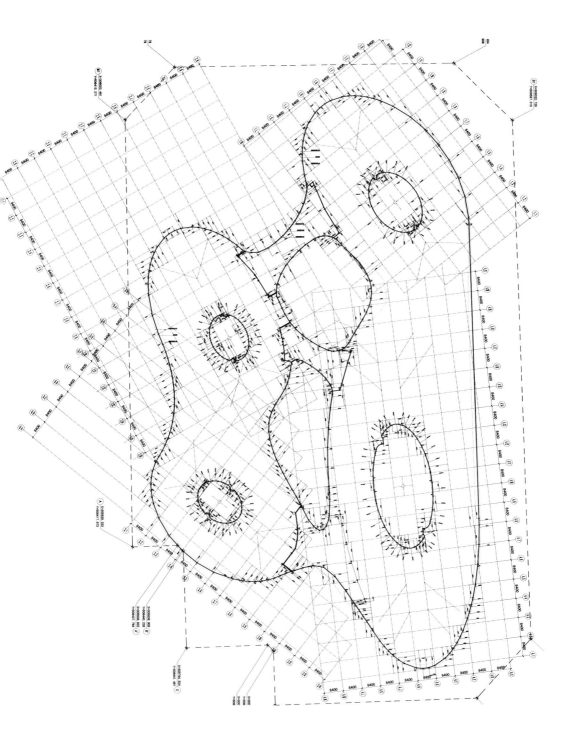

(top-left)Low Angle View at the atrium; (bottom-left)atrium; (right)slab edge set-out

序列

Sequence

4.18

设计时间：
2012
完成时间：
2012

威尼斯建筑双年展是国际上最重要的建筑艺术展览，每两年举办一次，因其先锋态势，一直是建筑艺术和展览潮流的风向标。2012年第13届威尼斯建筑双年展的总策展人是英国建筑师大卫·奇普菲尔德，主题是"共同基础"，热切倡导双年展更多地关注建筑文化的连续性、文脉和记忆，强调共同的影响和期待。邵韦平先生的展示作品《序列》作为中国国家馆5个参展作品之一，于2012年8月27日在威尼斯建筑双年展中国馆的开幕式上精彩亮相。《序列》作品以凤凰中心项目为概念原型，沿一个隐性的建筑环形轴网切出96个剖面，再将其沿直线按顺序展开，形成了一个23米长的序列阵，并通过量身定制的灯光效果塑造出了一条在空中"舞动的龙"。这一构思和再创作过程贴切地回应了中国馆策展人方振宁提出的"原初"的策展主题，同时也以另一种方式重新解读和展示了建筑的空间形态。

（下图：切片概念分析图）

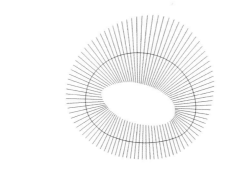

Design Time:
2012
Completion Time:
2012

The Venice Biennale of Architecture is the most important international architectural exhibition in the world. Held every two years, it has established itself as a barometer of architectural art and exhibition trends with its avant-garde posture. The theme of the 13th Biennale of Architecture in 2012 curated by British architect David Chipperfield, was "Common Ground", which advocated the Biennale to take more consideration of the continuity, history, and memory of architectural culture, and emphasizing common influences and expectations. As one of the five exhibition works of the Chinese Pavilion, Sequence by Mr. Shao Weiping was shown at the opening ceremony of the Chinese Pavilion at the Biennale on August 27, 2012. The concept prototype of Sequence comes from the Phoenix Center project. It cuts out 96 sections along the invisible architectural annular axis network, then unfolds them along straight lines, forming a 23-meter-long sequence array, which creates a "dancing Chinese dragon" in the air using the customized lighting effect. Such a conception and recreation process perfectly responded to the curatorial theme of "Originaire" proposed by Fang Zhenning, the curator of the Chinese Pavilion. It also reinterpreted and demonstrated the spatial form of the architecture in another way. (Above: Slice conceptual analysis)

数字技术的应用，在以凤凰中心项目为原型的展示作品中得到了延续。数字技术的介入，使该作品具备了更整体和更精良的视觉效果，同时展现了更鲜明的时代特征。

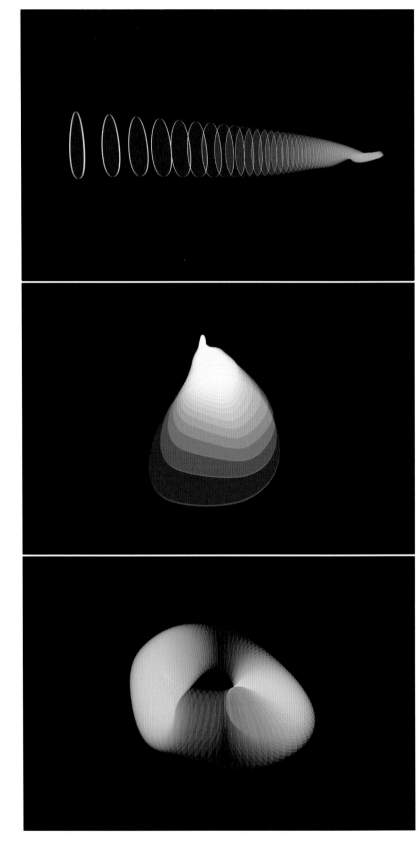

In this exhibition work that takes Phoenix Center project as the prototype, digital technology application not only continues, but also endows the work with a more integral and more superior visual effects, and reveals more distinct characteristics of our time.

(left)3D model; (right)at the exhibition

城市化

CBD核心区公共空间 [4.21]
奥体文化商务园区公共空间 [4.22]
北京国际图书城 [4.23]
北京世纪华侨城 [4.24]
华侨城北京总部 [4.25]
北京华侨城社区学校 [4.26]
中石油总部 [4.27]
朝阳区规划展览馆 [4.28]

Urbanizaiton

CBD Core Area Public Space [4.21]
Olympic South: Culture Zone, Business Park and Public Space [4.22]
Beijing International Book Mall [4.23]
Beijing OCT [4.24]
HQ of OCT Group Beijing [4.25]
School of Beijing OCT [4.26]
Headqarter of China National Petroleum Corporation [4.27]
Chaoyang District Urban Planning Exhibition Hall [4.28]

城市化:城市建造 / 刘宇光

UFo从建立之初,一个主要工作方向就是研究北京城市空间发展问题,希望建立从城市空间的整体架构,到区域空间的组织,再到局部建筑空间的一体化脉络。

北京城市空间的基本尺度,通常是由城市干道、快捷道路或主干道围合成的街区,大小在500米×500米左右,远远超过欧美城市传统街区的尺度。这种划分方法并不是基于最经济的土地利用模式。在这个框架内,土地由一个或多个土地产权单位所占用,也有内部道路,这个尺度我们可以称之为北京的街区尺度,它们像细胞一样填充了城市的总图。街区内部的产权单位有各自的道路系统,但各不相连,建筑及空间的组织都是各自独立存在。沿街区外侧排列着主要公共建筑,街区内部空间则比较封闭。这样的街区是我们研究的对象,包括现有街区的调查研究和新建街区的设计研究。研究的目的不是为了回归传统城市,而是希望在这样的尺度上,建立起以步行驱动的人性化活动空间,创造功能更加混合、空间更加开放的新型城市单元,从而激发城市活力,提高土地及空间效率,尤其是有机会进行整体规划和建设,使建筑更多地和其所在的场所相关联,和城市基础设施相关联。

UFo从2003年开始进行北京华侨城的总体设计,目标是要建造一个集居住、旅游、商业、演艺、教育、办公等多种城市生活为一体的小型城市片区。通过10年的全过程工作,我们完成了区域内所有类型的建筑项目的设计,已经实现了当初的规划理念,形成了持续发展的多元混合的城市活力区。

Urbanization: Urban Construction / Liu Yuguang

One of UFo's main focos is the spatial development issues of Beijing, with a view to establish an integrated sequence from the overall structure of urban space to the organization of regional space, and to the partial architectural space.

The basic scale of Beijing urban space is generally constituted by blocks enclosed by arterial streets, shortcut roads or main streets. The dimension is about 500m x 500m, far surpassing the scale of conventional blocks in Western cities. Such a partition method is not based on the most economic land use. Within this framework, the land is occupied by one or multiple property units with internal streets. Such a scale may be called the block scale of Beijing, filling up the city's map like cells. The property units of internal blocks have their own road systems, but with no connection to each other. Organization of buildings and spaces exists separately. For each block, the external space is lined with main public buildings, while the internal space is relatively closed. Such blocks are our research object, including the studies of existing blocks and the design and research of newly built blocks. The objective is not to regress to the conventional city but to establish a human scale activity space driven by walking, and to create new city units with more hybrid functions and more open spaces, to stimulate urban vitality and improve land and space utilization efficiency. In particular, it will be possible to conduct integrated planning and construction to make the buildings more connected with the context they are in and with urban infrastructure.

The design of Beijing OCT initiated in 2003 is to establish a small urban area integrating various urban lives such as living, tourism, business, artistic activities, education, and offices. Through ten years of whole process work, we've completed designs of all types of buildings in the area, fulfilled the original planning concept, and successfully shaped a sustainable, diversified urban area.

2005年开始的北京国际图书城项目，业主要求在两年时间内完成一个超大型综合体，包括亚洲最大的书库、国际图书交易中心、配套酒店、宿舍等内容。我们策略性地采用模数控制、工厂预制机电集成等方法，配合严格的过程控制，不仅完成了业主这提出的几乎不可能完成的任务，也实现了建筑师的意图，使这个项目成为快速城市化发展的一个典型案例。

2008年奥运会以后，北京的城市建设进入了相对稳定的阶段，城市格局基本成型，城市基础设施陆续建成。城市品质的提升吸引了更多的产业和人口，结果导致环境恶化、交通拥堵等城市问题，让城市空间迅速达到饱和状态。由于城市不可能无限制地向外扩张，因此提高土地使用效率，精细化利用城市空间，就成为城市发展的重心。追求效率不是非人性化的，目的要使城市、建筑和人能和谐相处。经过对北京城市构架的研究，我们认为网格化街区这个城市最基本的尺度，在以往的用地规划中受到很多制约，在建筑设计中也被低估了，没有发挥应有的作用。

我们从2008年开始做北京CBD核心区和奥林匹克中心区南区两个新建街区，尺度是典型的北京街区尺度，但强度、密度和高度都是前所未有的。我们所进行的街区设计主要包括四个方面：街区总体设计、框架设计、公共空间设计，以及地块建筑设计。

街区设计不同于建筑综合体的设计，强调的不是形式和结果，而是秩序和规则。恰如为建筑和城市找一个中介，更加强调开放和多元性，因此具有更长久的生命力。对城市而言，每个街区都有不同的涵义，这种基本框架既可以打破千城一面的呆板局面，又能保证城市品质。

For the Beijing International Book Mall project started from 2005, the client requested an ultra-large complex including the largest stack room in Asia, international books trading center, hotels, etc., be completed within two years. We strategically adopted methods such as modulus control and prefabricated electro-mechanical integration, along with strict process control to finish the almost impossible mission, and to realize the intention of the architect. It has become a model case of rapid urbanization.

Affter the wave of the 2008 Beijing Olympics, urban construction entered a relatively stable stage. The city pattern was generally shaped and urban infrastructure was successively built. Improvement of city quality attracted more industries and population, resulting in problems such as environmental degradation and traffic jams. Urban space has rapidly been saturated. Since the city cannot expand without limit, to improve land utilization efficiency and smartly use the urban space becomes the focus of city development. The pursuance of efficiency is by no means dehumanized, but aims at the harmonious coexistence of city, buildings, and people. As a result of researching urban structures of Beijing, we believe that the grid block, the basic scale of the city, was much restricted during the previous land utilization planning, and underestimated during the construction design. It did not work as efficiently as it should have.

We have been working on two new blocks in Beijing CBD core area and the southern part of the Olympics central area since 2008. The scales involve typical Beijing blocks, but the strength, density, and height are unprecedented. Our block design includes four parts: overall design, framework design, public space design, and architectural design.

Different from the design of a complex, to design a block does not emphasize form and results, but order and rules. It's like finding an intermediary between buildings and city, therefore places more emphasis on openness and diversity, and hence more enduring. The meaning of blocks varies with cities. The basic framework can break the same rigid image of cities while securing the quality of urban life.

CBD核心区公共空间

CBD Core Area Public Space

4.21

用地面积:
113 172平方米
建筑面积:
524 130平方米
建筑高度:
10米
设计时间:
2003—2011
建造时间:
2011—

北京CBD核心区公共空间及市政交通基础设施项目是核心区总体建设的奠基石,以50万平方米的规模,为核心区内每一幢建筑提供方便快捷的城市交通联系、完善的市政配套设施和应急防灾避难空间;地面上的中央广场和下沉花园具有公共活动、观光休闲,疏导交通和商业服务的功能。本工程不单是一个地下工程,而是集公共景观、道路市政、建筑空间三位一体的高品质新型城市综合体代表了CBD规划建设的品质。(下图左:东西组团功能性差异;右:标志性建筑与公共空间的对位)

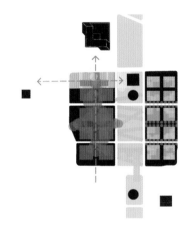 

Site Area:
113,172m²
GFA:
524,130m²
Building Height:
10m
Design Time:
2003-2011
Construction Time:
2011-

Beijing CBD core area public space and municipal traffic infrastructure project is the cornerstone for the overall construction of the area. It covers an area of 500,000 square meters, offering convenient and fast urban transportation, complete municipal facilities and emergency shelters for every building within the core area. The above-ground central square and sunken garden function as spaces for public activities, sightseeing and recreation, traffic dispersion, and business services. This project is not only an underground project, but also represents a new type of high-quality urban complex that integrates public landscape, municipal roads, and architectural space. It demonstrates the quality of the planning and construction of CBD. (Above left:function comparison of the west and the east; right: landmarks and public space on the bit)

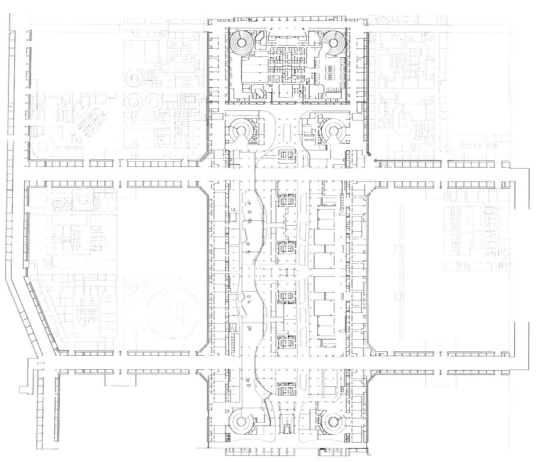

城市主板：在交通系统的基础上形成了一个整体性的30公顷的"城市主板"，它促进了城市和建筑的接驳，使18个超高层建筑像插件一样，可以和城市进行对接，形成一个完整的城市商务街区。这个街区体现出集约化、高密度化的趋势，是一次关于立体城市的探索。

城市框架：利用城市道路建立立体的"型材"，将人行交通、车行交通、市政管线和综合管廊分层立体设置，并且综合建筑垂直交通和进排风竖井，形成完整的竖井体系。城市框架是连接城市市政基础设施与建筑空间的尝试。

City motherboard: An integral "city motherboard" of 30 hectares emerges on the basis of the traffic system, which promotes the connection between city and architecture, so that 18 skyscrapers, like plug-in units, can be connected to the city to form a complete business block. A trend of intensive high density is reflected in this block, which is an exploration towards a three-dimensional city.

Urban framework: Three-dimensional "profiles" are established using urban roads, which are arranged by layers of pedestrian traffic, vehicular traffic, municipal pipelines and a composite pipe gallery. Vertical traffic and inlet and exhaust shafts are integratedly constructed, forming a complete shaft system. The urban framework is an attempt to connect municipal infrastructure and architectural space.

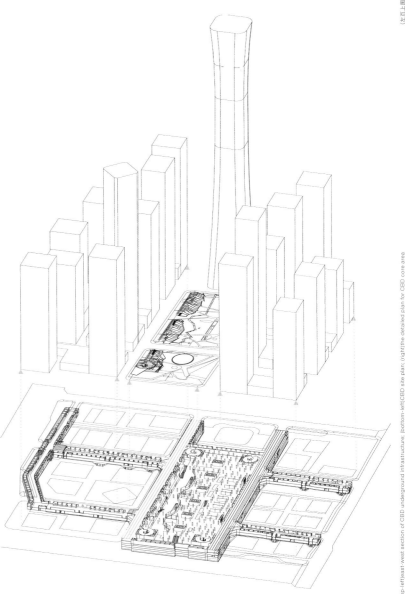

（左页上图）核心区地下空间东西向剖面图；（左页下图）CBD总平面图；（右页）CBD核心区总规划

(top-left)(east-west section of CBD underground infrastructure; (bottom-left)CBD site plan; (right)the detailed plan for CBD core area

165

地下空间：地下空间共分五层，包含了人行交通、商业服务、停车及安全避难功能，地面是核心区的公共绿地，形成了50万平方米的城市公共空间，规模庞大的绿色基础支持了超高强度的地面开发。

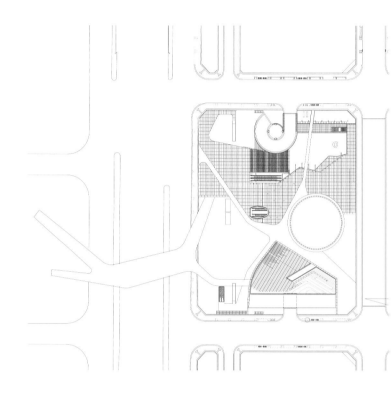

Underground space: It includes five floors, which contain pedestrian traffic, business services, parking, safety and evacuation functions. Above the ground is the green space, forming an urban public space of 500,000 square meters, and the large green foundation supports the intense development on the ground.

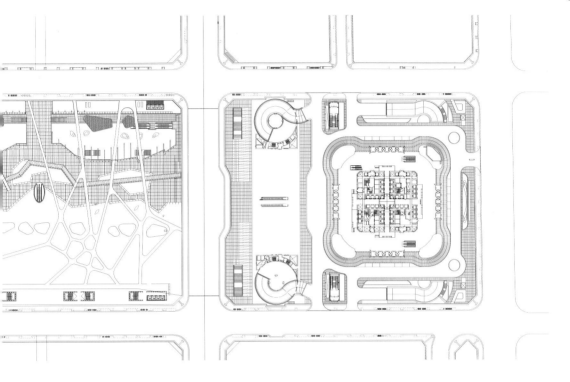

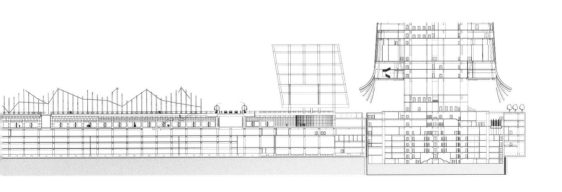

(top)ground floor plan; (bottom) cross sectional view

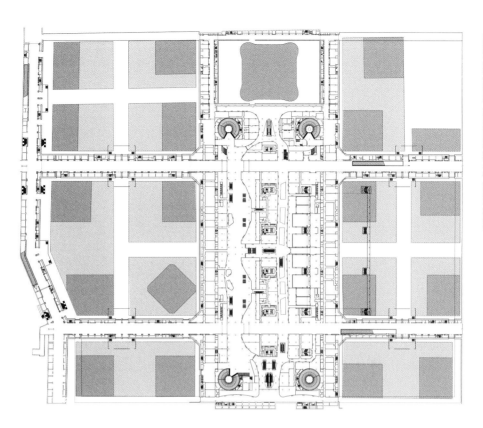

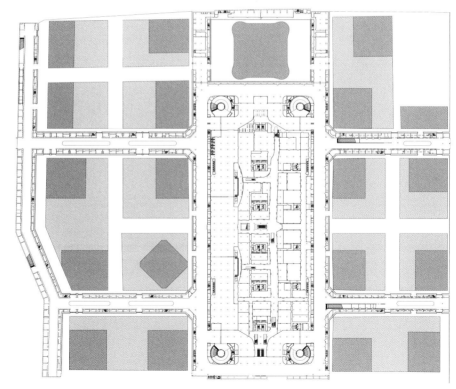

(left)during construction; (top-right)B2 plan; (bottom-right)B1 plan　0———50m

（左页图）施工过程；（右页上图）地下二层平面图；（右页下图）地下一层平面图

奥体文化商务园区公共空间

Olympic South: Culture Zone, Business Park and Public Space

4.22

用地面积:
281 869.9平方米
建筑面积:
300 000平方米
建筑高度:
7米
设计时间:
2011—2014
建造时间:
2012

奥体文化商务园区公共空间总规模约20万平方米，地面上的中央广场和下沉花园具有公共活动、观光休闲、疏导交通和商业服务的功能；通过对奥体南区周边业态的分析，希望能在此区域创造一种包含商业办公、文化设施、体育设计、公寓酒店等多样功能混合的城市功能业态。地下利用北辰东路隧道进入园区的有利条件，在地下二层设计内部环隧，直接入口连通各地块车库，实现净化地面交通、缓解亚北地面交通的目的。采用局部道路立交的方式，将二号路抬高七米，形成东西向车行主路，并且采用景观地形的方法，使中心绿地与道路缓坡保持一致，联系东西向人行空间，形成独具特色的"新奥湾"中央景观，在隧道口部上方增设架高人行平台，连接东西南北轴线，增加与奥体中心活动空间的联系。

(下图: 总平面图)

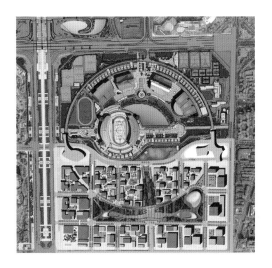

Site Area:
281,869.9m²
GFA:
300,000m²
Building Height:
7m
Design Time:
2011-2014
Construction Time:
2012

The public space of Olympic Culture and Business Park covers an area of 200,000 square meters. The above-ground central square and sunken garden function as spaces for public activities, sightseeing and recreation, traffic dispersion, and business services. Taking advantage of the East Beichen Road Tunnel extending into the underground area of the Culture and Business Park, we set an internal ring tunnel on the second floor underground, which connects up the underground parking lots at different land lots, and reduces pressure on ground traffic to the north of the Asian Games Village. We also make use of partial Road Interchange and lifted the No. 2 Road by 7 meters, making it a major east-west main roadway. At the same time, we utilize techniques of landscape design to reconcile the central greenland with the gentle slope of the road and establish an east-west pedestrian corridor, forming a unique "New Olympic Bay" central landscape. Furthermore, an elevated pedestrian platform is added above the tunnel entrance, which connects the east-west axis and the north-south axis, and improves the connection between the Park and the Olympic Sports Center.

(Above: site plan)

中央景观绿地采用人工地形后,坡地下面的空间就可以充分利用,其中部分空间为人行和地铁公交站点换乘创造了人行主街,为周边地块提供人行支持;部分开敞空间用作公共服务酒店,提升地区服务品质;在其下方结合人防消防体系,设置综合避难场所、集中能源机房和综合管廊。

立体化的道路交通网络
园区内市政红线范围内设置地下环形隧道,布置在地下二层高度,串联园区地块内地下二层车库,行驶方向为逆时针,同时环形隧道与北辰东路南延隧道通过地下定向匝道连接。园区内提倡公共交通,其中有两个地铁站和一个公交首末站,因此形成了布局清晰、层次分明的三级人行体系。

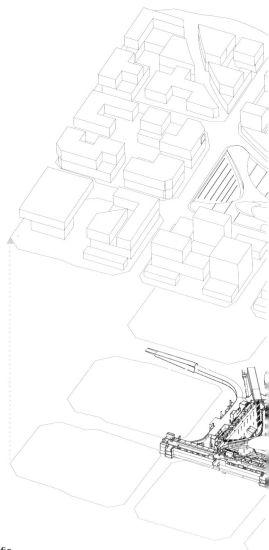

As artificial topography is adopted for the central landscape greenland, space under the slope can be made full use of: part of the space is planned as a major pedestrian street leading to subway stations or bus stops, which also supports the pedestrian traffic of neighboring land lots; partial open space is utilized to build hotels in order to boost service quality in this district. Beneath it are comprehensive emergency shelters that integrate air-defense and firefighting systems, a central energy machine room and a comprehensive pipe gallery.

Three-dimensional road traffic network
With an underground annular tunnel set up on the second floor underground, within the scope of the municipal red line in the Park, the second-floor garage underground within the park block is connected in the counterclockwise driving direction. Meanwhile, the annular tunnel is connected to Nanyan Tunnel on East Beichen Road through an underground directional ramp. Public transportation is encouraged in the Park. Two subway stations and a bus terminal are included in it, thus forming a three-level pedestrian system with a clear layout and distinct levels.

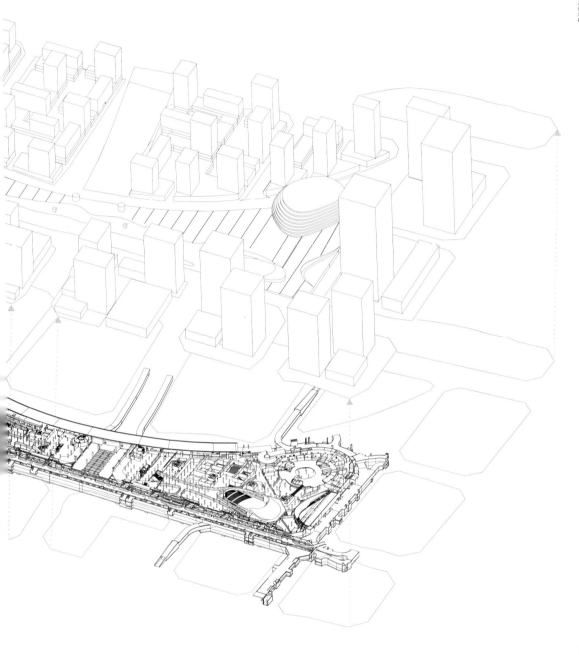

the overall model of olympic culture business park 奥林文化商务园区总体模型

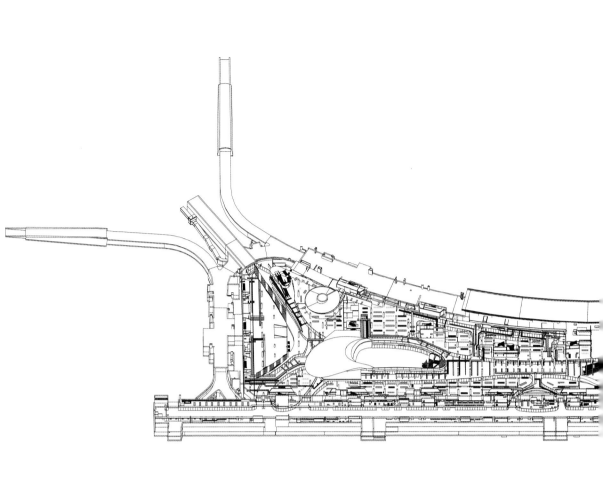
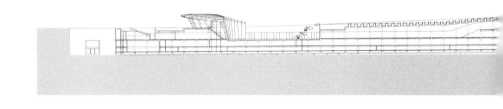

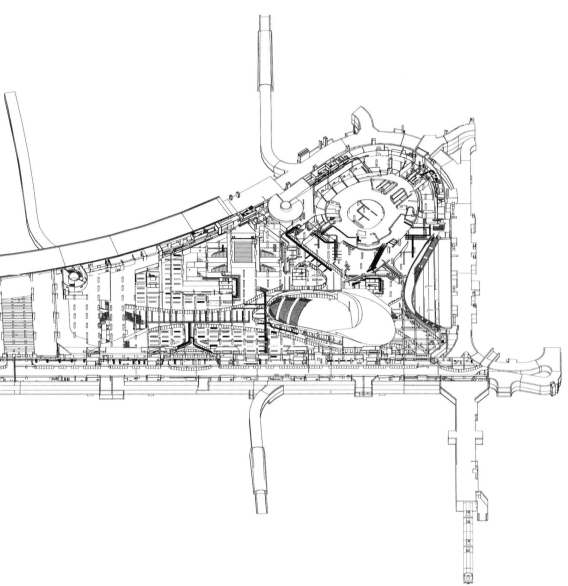

地下一层BIM模型

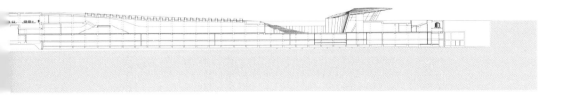

bim models basement

都市中的慢坡绿地
城市森林、城市客厅和下沉花园的穿插组合，能够烘托周边高层建筑的挺拔壮丽，激发空间的活力，同时营造出人性化的生态环境。

一体化的市政能源综合沟
在市政道路红线下，地下一层和地下三层分别设置市政一级和区域二级综合管廊。
综合管廊的设置不但保持了市政道路的完整性，还可以降低因管线维修施工造成的路面翻修费用。由于利用的是市政道路地下空间，因此也节省了城市用地，方便日后的运行与管理。

Gentle slope and green space in the city
Interspersion and combination of an urban forest, urban living room, and sunken garden sets off and contrasts the tallness and magnificence of surrounding skyscrapers, stimulates the vitality of the space, and creates a humanized ecological environment.

Integrated municipal energy integral pipe trench
Under the municipal road red line, the first and third floors underground are set up with a composite pipe gallery of municipal level 1 and regional level 2 respectively. Setting up the comprehensive pipe gallery will not only maintain the integrity of municipal roads, but also reduce pavement renovation costs caused by pipeline maintenance. Meanwhile, using the underground space of municipal roads will save urban land, and facilitate their operation and management in the future.

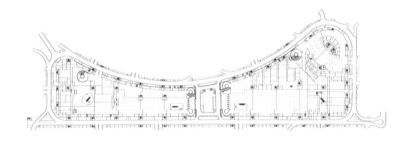

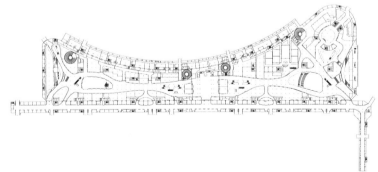

(top-left) during construction; (middle-left) B1 plan; (bottom-left) B2 plan; (top-right) main street business; (bottom-right) ramp　　0 ——— 50m

（左页上图）施工过程；（左页中图）地下一层平面图；（左页下图）地下二层平面图；（右页上图）商业主街；（右页下图）环形坡道

北京国际图书城

Beijing International Book Mall

4.23

用地面积:
241 000平方米
建筑面积:
76 300平方米
容积率:
3.4
建筑高度:
23米
设计时间:
2005—2006
建造时间:
2006—2007

北京出版发行物流中心位于北京市东郊通州区台湖镇,占地面积24.1公顷,一期建筑面积242 071平方米,包括物流仓储配送中心、出版物展销中心、配套服务中心和接待中心,是全国最大的出版物集散中心。规划采用简洁明确的布局方式,将建筑、环境、人流、货流统一考虑,整合为四组长条形区域,以应对该项目的复杂功能。从西向东,整个地块被划分为仓储配送区、高效物流通道、展销贸易区、绿化景观带。整个园区10栋建筑从设计到施工历时一年九个月落成开业,成为中国式速度的又一例证。
北京国际图书城是北京出版发行物流中心的主楼。书城以5.7万平方米的图书买场面积成为全球最大书店。这座现在被叫作"8字楼"的建筑,整体空间组织以数字"8"作为设计图解:用一个在平面上贯通南北两端、高度上贯通三层的中庭连接每层的四个展销厅,连廊和台阶在中部和南北两端将各层展销厅从平面和剖面上连接起来。由此,全部展厅在空间上构成一个循环连续的"8"字形,产生了丰富流畅的内部动线和简单强烈的整体建筑形象,并且充分贯彻到细部设计中。(下图: 流线模型)

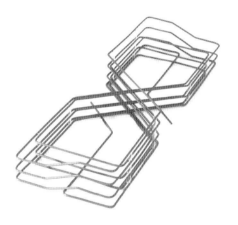

Site Area:
241,000m²
GFA:
76,300m²
FAR:
3.4
Building Height:
23m
Design Time:
2005-2006
Construction Time:
2006-2007

Beijing Publication Logistics Center is located in Taihu of Tongzhou District, an eastern suburb of Beijing. It occupies 24.1 hectares. The floor area of the first phase project is 24,2071m². Consisting of the Logistics Storage and Distribution Center, Publication Exhibition Center, Service Center and Reception Center, it is the biggest publication collecting and distributing center in China. The succinct and explicit layout is adopted in the design, taking the buildings, environment, people and goods into general consideration. Four groups of long strip areas are integrated in order to fulfill the complicated functions of the project. From west to east, the whole plot is divided into a storage and distribution area, high efficiency logistic channels, an exhibition trade area and a green landscape area. It took one year and nine months for all ten buildings to complete, from design to construction and realization. Another good example of Chinese speed. Beijing International Book Mall is the main building of the Logistics Center. It is the biggest bookstore in the world, covering 57,000 km² trading area. The 8-shaped building is designed in the figure "8" as a diagram in the whole space organization. That is, by applying an atrium that horizontally crosses the north and south and vertically runs three floors, connecting every four exhibition halls on every floor. Corridors and staircases in the middle and the end of north and south link the exhibition halls on every floor, both horizontally and vertically. Therefore, all exhibition halls form a consecutive circle like "8" in space, which builds a fluent inner line as well as a simple and strong image for the whole building. It also can be seen in the detailed design. (Above: Flow line model)

数字"8"作为图解直接产生了设计概念形象，1：2000的体块模型固化了这一设计概念，是从流线图解到建筑形体的决定性转换。概念的清晰准确使得最终在使用中，"8字楼"取代了正式名称而成为日常称呼。

为了梳理建筑的内部动线，我们用一根三米长的铁丝弯成一个流线模型（见前页图），这一制作方式本身就直接反映了建筑内部动线特征：同层及邻层连接，使整个流线构成一个循环连续的"8"字形。

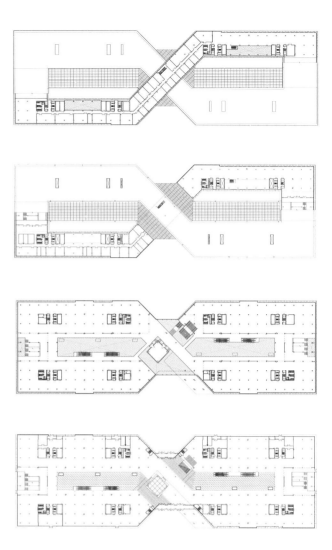

The figure "8" was used as a diagram and directly generated a design concept, which then was solidified by a 1:2000 model. This was a decisive transformation from a circulation diagram to an architectural form. Due to the clarity and accuracy of the concept, the "8-like building" eventually becomes a usual name in place of the official one.

To figure out the internal flow line, we used a 3-meter long iron wire and bent it into a circulation model (previous page) – the way it was produced directly reflects the internal circulation within the building: the same floor connection and adjacent floors connection create a successive "8" on the whole circulation.

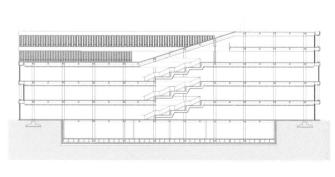

在形体和空间研究阶段,针对不同的问题,我们制作了大量不同比例的模型,其中1:500和1:300楼层模型主要研究建筑的空间组织方式。模型中,由于制作材料使用了五毫米厚的KAPA板,其效果也促成了最终一米高素混凝土外挑板的典型细部处理。

During the research phase on structure and space, in light of different problems, we have made a great many models of different proportions, among which 1:500 and 1:300 storey models mainly focus on spatial organization of a building. Among the models, because we use 5mm KAPA plane as material, we have finally achieved such a typical detail treatment as 1m concrete overhanging plates.

竖向百叶给建筑形式带来流动感,其排布也遵照着严密的组织逻辑:上下楼层类型相异;平面东西边斜边对外,南北边尖角对外;转角类型相同。这一研究达到了预期效果:两类百叶连续排布在相邻楼层间,呈现出的差异与外挑的素混凝土挑板一起,给立面效果带来了强烈的流动感,消解了看似对称的总体布局。

设计和施工过程中,我们制作了大量实体模型。对立面竖向百叶的截面形式和排布方式进行了充分研究,通过大量1:1百叶单元模型的对比,最终将百叶截面形式确定为窄长的平行四边形。它切合平面轮廓角,而且通过镜像使用获得了两个方向性,为其后的整体排布方式研究提供了有力的基本元素。

Vertical lovvre bring about a feeling of flow to architectural form; their configuration is controlled by strict logical organization: different types for upper and lower floors; in the plan, slope sides facing outwards in the east-west direction, and sharp edges outwards in the north-south direction; same type of corners. This study has met the expectation: two kinds of shutters are of a continuous arrangement between adjacent floors, and the differences presented along with the projecting plain concrete slab bring a strong dynamic feeling to the facade, dispelling the seemingly symmetrical general layout.

A large number of models were produced in the design and construction process. Adequate research on section forms and arrangement of the vertical façade shutters were performed, and through comparison of a large number of 1:1 lovvre unit models, the section form of shutters was eventually determined as a long, narrow parallelogram: it suits the contour angle and obtains two directions by mirroring, providing powerful basic elements for further study of the overall arrangement.

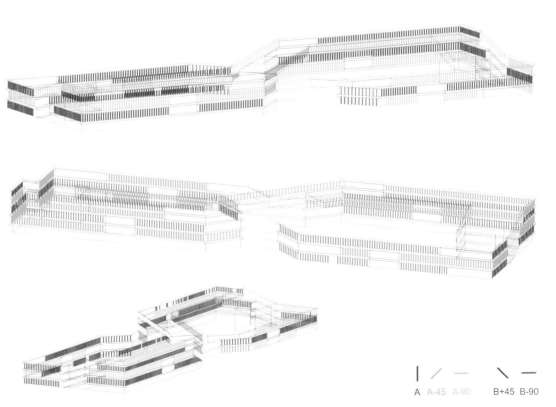

A A-45 A-90   B+45 B-90

(left) view from southeast; (top-right) atrium; (bottom-right) Louver type research

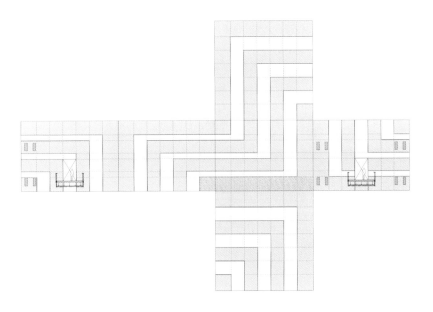

北京世纪华侨城

Beijing OCT

4.24

用地面积:
345 200平方米
建筑面积:
350 000平方米
容积率:
2.43
建筑高度:
47米
设计时间:
2003—2004
建造时间:
2004—2005

位于北京东四环的世纪华侨城经过十年的规划和营建,已经由郊区的村庄变成了一个乌托邦式的小城。它包含了居住、城市公园、体验式商业综合体、学校、剧院、办公等很多内容,几乎涉及当代生活的各个方面。伴随着从规划到实施的全过程,设计的画面变成真实的场景,我们也在不断观察、思考、校正设计和真实的城市生活之间的距离,设计师的意志与人的活动如何保持关联。在中国快速城市化过程中,北京华侨城是一个很典型的缩影。(下图: 总平面图)

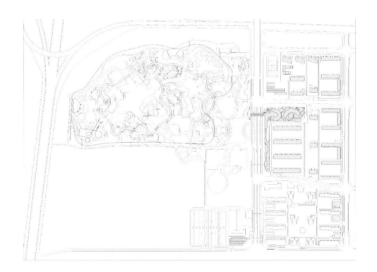

Site Area:
345,200m²
GFA:
350,000m²
FAR:
2.43
Building Height:
47m
Design Time:
2003-2004
Construction Time:
2004-2005

Located on the East 4th Ring Road, Beijing Century Overseas Chinese Town (Beijing OCT), after 10 years of planning and construction, has become an ideal town from a suburban village. It includes housing, urban parks, an experiential commercial complex, schools, theaters, offices and so on, covering almost all aspects of contemporary life. With the whole process from planning to implementation, design pictures have turned into a real scenario. Meanwhile, we have been observing, thinking, and correcting the gap between design and real urban life, and dealing with the association of the designer's will and human activities. In the process of rapid urbanization in China, Beijing OCT is an epitome of the process. (Above: site plan)

华侨城北京总部

HQ of OCT Group Beijing

4.25

用地面积:
3 004平方米
建筑面积:
4 282平方米
容积率:
1.4
建筑高度:
9.3米
设计时间:
2005
建造时间:
2005—2006

该项目是华侨城集团内部使用的办公楼,业主希望用低造价达到理想的效果。为此,我们试图使用最简单的手法去解决问题。为了打破通常板式办公楼的单调形式,设计采用了"取长补短"的基本概念:将二层楼两个柱网的体块切下旋转移至南侧外,形成八米悬挑体块。这一动作获得了以下功能:二层东侧独立的领导办公区,两个办公区之间的屋顶花园,悬挑的大会议室,悬挑部分也成为其下方入口的雨篷。(下图:设计概念分析图)

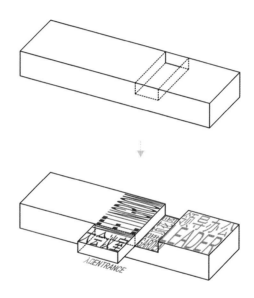

Site Area:
3,004m²
GFA:
4,282m²
FAR:
1.4
Building Height:
9.3m
Design Time:
2005
Construction Time:
2005-2006

The project is the HQ Building of Overseas Chinese Town. The owner requested an ideal effect with economical cost. We used the simplest method to respond. In order to break through the monotonous form of ordinary slab-type office buildings, the basic concept of "using strong points to offset weakness" was adopted in the design - two column grid blocks on the second floor were cut off and shifted to the south side of the building, forming an 8 meter cantiliver block. The following functions were achieved by this move: a separate management office area on the east side of the second floor; a roof garden between two office areas; an overhung large conference room which also functions as the awning for the entrance beneath. (Above:design concept analysis)

结构悬挑：结构工程师的创造性努力使八米悬挑部位的结构高度控制在80厘米以内。

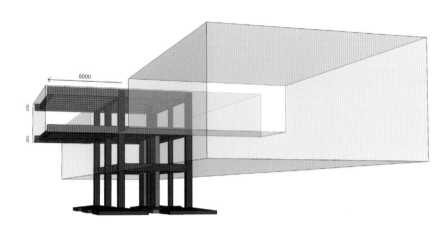

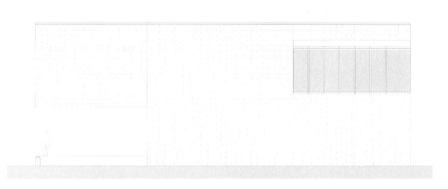

Structural engineers have controlled the structure height of the 8 meter cantiliver part within 80cm.

(top-left) structure analysis; (bottom-left) east elevation; (top-right) entrance; (bottom-right) view from east

北京华侨城社区学校

School of Beijing OCT

4.26

用地面积:
20 900平方米
建筑面积:
12 060.1平方米
容积率:
0.57
建筑高度:
20.6米
设计时间:
2006—2007
建造时间:
2008

作为华侨城社区附设的九年制学校,它共设36个班。设计中我们着重于对社会现象和设计之间关系的探讨。现实世界两组矛盾:应试教育中沉闷的教条与鲜活的现实;社会现实中体制要求的相同性与天然个性要求的差异性。建筑形式和空间的两组反差:教学楼外立面形式与内部空间的不一致——具有复杂图案和色彩的外立面似乎暗示了复杂的内部空间构成,但实际上内部是匀质化的教室单元,二者毫无关联,反差强烈。教学楼空间的反差:没有差别的教室六面白色,地面采用匀质化网格;走廊墙面、地面、天花使用不同的颜色,地面铺装采用非匀质化网格。教室门内外是完全不同的世界。形式生成系统与最终形式:四个班级构成一个年级,这是学校内部基本的社会结构单元。以此为出发点,进行形式规则设定,最终的形式是在此规则下生成的众多结局的一种可能。(下图:设计概念分析图)

Site Area:
20,900m²
GFA:
12,060.1m²
FAR:
0.57
Building Height:
20.6m
Design Time:
2006-2007
Construction Time:
2008

The project is a nine-year school with 36 classes. Exploration of relations between social phenomena and architectural design was our focus during the design process. Two groups of contradictions in reality: tedious dogma in exam-oriented education versus active and fresh reality; homogeneity required by social system versus differentiation required by natural individuality. Two groups of contrasts in architectural form and space: the variance between exterior and interior space – the exterior facade with complicated patterns and colors seems to imply intricate composition of interior space. However, the inside is just the homogeneous ordinary classroom unit. There is no relation between the two aspects, so the contrast is striking. Spatial contrast in the teaching building – the interiors of the classrooms are totally white with no differences, same grid is adopted on the floor for all classrooms. For the corridors however, various colors are applied to the wall, the ground and the ceiling, and non-homogeneous grid is used as the floor covering. The inside and outside of the classrooms are two different worlds. The form generation system and the final form: Four classes form a grade, which is the basic social unit inside the school. The form regulation setting is performed on this basis, and the final form is just one of the many possible results generated from such regulation. (Above: design concept analysis)

在学校教学楼立面设计中,我们设计了一个形式生成系统。我们选取学校内部的基本社会结构单元四个班级(一个年级)为出发点,进行形式规则设定,最终的形式是在此规则下生成的众多结局的一种可能。

具体步骤如下:a. 四个班可以形成七类图形,每一类中包括若干种图形;b. 测试这七类图形和立面划分的吻合程度,得到两个方向;c. 我们认为方向一极为单调死板,生成的形式很有限,所以选定方向二继续发展;d. 在方向二的指导下,生成数以百计的图案组合;e. 挑选令人愉悦的图案,以三维模型的方式进行测试并求业主意见;f. 业主反馈了意见后,我们再从图案库中挑选新的图案进行测试。这样的过程持续了三次后定稿。

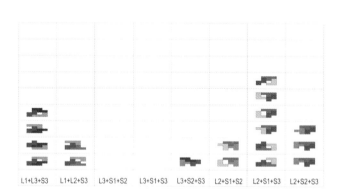

Module generation system was designed when we were thinking about the facade design for the school teaching building. One grade (i.e. four classes), the basic unit of social structure inside the school, served as a starting point for setting up rules for the form, and the final form is one possibility out of many resulting from the rules.
The method contains following steps:
The four classes can result in seven types of graphics, each type including several kinds of graphics.
Test the consistency degree between these seven types of graphics and the facade division, which result in two directions.

Direction I is considered as dull and rigid, which would only generate a limited amount of forms. So we decide on Direction II to proceed. Hundreds of pattern combinations are generated under the guidance of Direction II.
Certain pleasing patterns are picked and tested using 3D model; test results are provided to the client.
Upon the client's feedback, we choose new patterns to run test. This process is repeated three times until we get the final result.

(left)pattern diagram; (right)staircase
(左页图)形式系统图解；(右页图)楼梯

中石油总部

Headqarters of China National Petroleum Corporation

4.27

用地面积:
9 321平方米
建筑面积:
200 898平方米
容积率:
6.36
建筑高度:
90米
设计时间:
2006—2007
建造时间:
2008

中石油总部位于北京东直门东二环交通商务区北部入口,是著名能源企业的总部办公楼。建筑设计充分体现环保、节能的时代特点,用高技术手段实现可持续发展。中石油总部面临的问题是场地特别狭长,南北长335米,东西宽仅55~74米,是典型的街墙式建筑,而且场地西侧还有高层住宅,不能造成遮挡。我们利用了用地细长的特点,用组群化的布局,化不利为有利,以"L"形母题不断演进,顺应地形要求,产生主次关系。UFo作为设计团队之一完成了方案创作工作。(下图: 模型照片)

Site Area:
9,321m²
GFA:
200,898m²
FAR:
6.36
Building Height:
90m
Design Time:
2006-2007
Construction Time:
2008

China Petroleum Plaza, located at the north entrance of the Transport and Business District, East 2nd Ring Road, Dongzhimen, Beijing, is the headquarters building of the well-known energy giant. Its architectural design fully embodies the epochal characteristics of environmental protection and energy saving, with high-technology means to realize sustainable development. The challenge we faced was the site was particularly long and narrow: 335 meters long in the north-south direction, 55~74 meters wide - a typical street-wall building. Moreover, there are high-rise residencials to the west, which cannot be blocked from sunlight. Taking advantage of the long and narrow characteristics of the site, we used a clusterized layout to turn the disadvantage into an advantage, continuously evolving with an "L" motif, complying with the topographical requirements, and creating a primary-secondary relationship. UFo was responsible for the schematic design of the project. (Above: model photo)

对位关系：以中间双"L"形主楼对应东侧东直门交通枢纽，南配楼对应东直门立交桥，北配楼对应东二环北入口。有明显对位关系且"一体两翼"、主次分明的总图布局，充分利用了基地的长度，形成井然的建筑秩序。

视线走廊：四个相对的"L"形母题的有序演进，形成了丰富的建筑外部空间，且从二环主要视点形成视线的通透，一改沿主要路封闭街墙的呆板形态，或山墙向街的兵营式布局，使城市空间收放自如。

Counterpoint relationship
The double L-shaped main building in the middle is designed to correspond to the Dongzhimen traffic hub on the east. The southern wing corresponds to the Dongzhimen Flyover, and the northern wing to the north entrance to the East 2nd Ring Road. The general layout with obvious counterpoint relationship, "one body and two wings," and distinctive primary and secondary features, takes full use of the site length, forming an orderly architectural sequence.

Visual corridor
The orderly evolution of 4 face-to-face "L" motifs forms an ample outer space of the building, and the unblocked sight from the main viewpoint on the 2nd Ring Road changes the rigid form of the closed street-wall along the main road, or the barrack style layout with the gables facing the street, thus enabling the urban space to breathe freely.

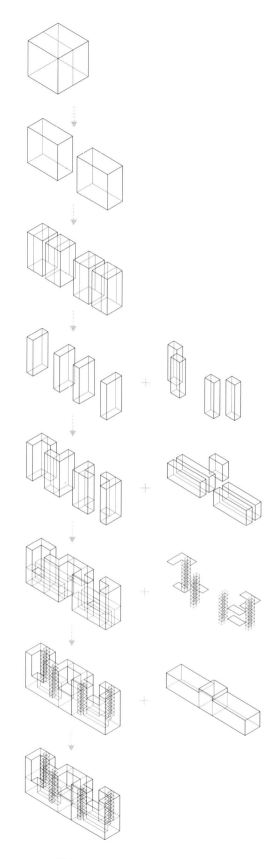

城市界面：长达248米的城市临街面，总图布局强调城市界面在规则中的变化，在稳重中的通透，使建筑以丰富有序的空间形态加入城市环境。

整体功能：248米的大厅贯通基地南北两大集群：北部为集团公司，南部为股份有限公司，分别位于东二环北入口和东直门立交桥，两个集群之间的核心区域是集团领导办公区。

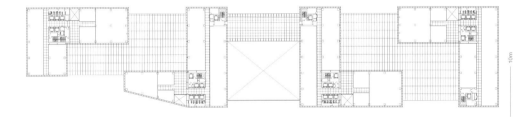

Urban interface
The general layout of a 248-meter long street facade emphaAreas changes in the well defined urban interface, and transparency in stability, enabling the architecture to join the urban space with an abundant and orderly form.

Overall function
The 248-meter long lobby connects the two large clusters on the site, with the northern for the Group Company and the southern part for the Limited Company, respectively showing up at the north entrance on the East 2nd Ring Road and Dongzhimen Flyover. The core area between the two clusters is office area for management group.

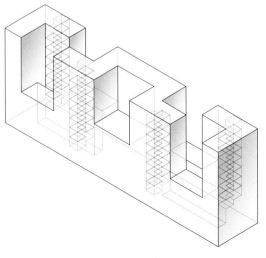

朝阳区规划展览馆

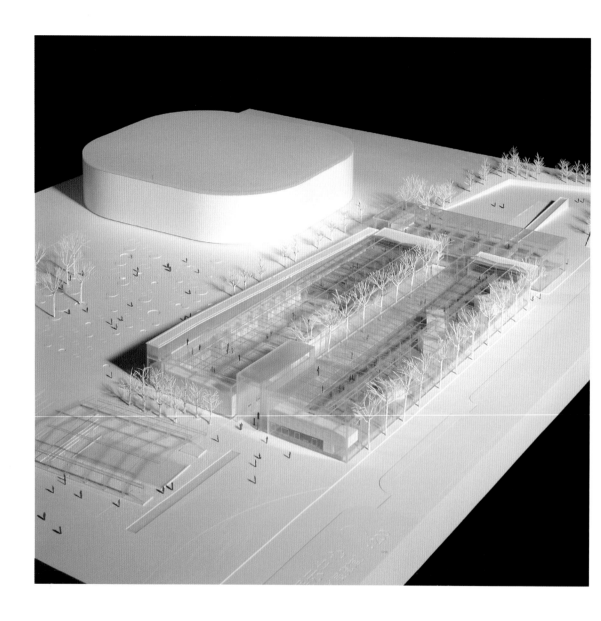

Chaoyang District Urban Planning Exhibition Hall

4.28

用地面积:
9 967平方米
建筑面积:
13 000平方米
容积率:
0.97
建筑高度:
10.8米
设计时间:
2009

北京朝阳展示中心由朝阳公园内原有旧厂房改造而成,位于公园北区东部,2008年北京奥运会沙滩排球场南侧。展陈是一座展览馆的灵魂。朝阳区城市展示中心的展示内容既有别于798的创意画廊,又不同于普通的以展示图片和模型为主的规划展览馆。如果简单套用目前国内常用的基本陈列、专题陈列和临时陈列并置的方式,由于题材所限,基本陈列会造成"一次性消费",临时陈列又受面积局限,难以满足高端展览的需求。这样的展馆往往难以为继。在我们的提案中,整个展示中心由一条巨大而连续的坡道构成。所有的主要展示空间都被这个独特的连续空间连为一体。(下图:活动区域分析图)

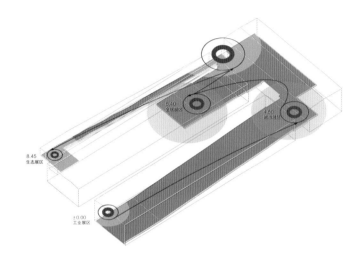

Site Area:
9,967m²
GFA:
13,000m²
FAR:
0.97
Building Height:
10.8m
Design Time:
2009

Beijing Chaoyang Exhibition Center is transformed from an original factory building in Chaoyang Park, located on the north east part of Chaoyang Park, south side of the Beach Volleyball Court for the 2008 Beijing Olympic Games. Exhibits are the soul of an exhibition hall. Items exhibited in Beijing Chaoyang Exhibition Center are different from either those in 798 art galleries or those pictures and models that an ordinary planning exhibition hall would feature. If we simply copied the current commonly used method of juxtaposing the general exhibition, feature exhibition and temporary exhibition, due to the limitation of subjects, general exhibition would turn out to be a "one-time consumption," meanwhile, subject to the space limitation, temporary exhibition would be difficult to meet the needs of high-end exhibitions. The Center would be hard to sustain in this way. Instead, we proposed a huge and continuous slope that will connect all the main exhibition spaces. (Above: Activity area analysis)

漫步于这个大楼梯,既是参观展览的过程,也是重温工业遗迹和奥运遗产的过程;既是艺术欣赏,也是休闲娱乐。朝阳城市展示中心具有主题展示与多功能活动相结合的特点,是一座全新概念的文化综合体,它将与摩天轮、莫比乌斯环一起,成为朝阳公园中具有吸引力的公共活动空间。

Walking along the gigantic stairway is not only a process of visiting the exhibition, but also a process of revisiting industrial sites and Olympic heritage. It is about art appreciation, alao about leisure and recreation. Chaoyang Exhibition Center has the characteristics of a theme exhibition combined with multi-functional activities, a cultural complex with a novel concept. Together with the Ferris wheel and Mobius ring, it will become an attractive public space in Chaoyang Park.

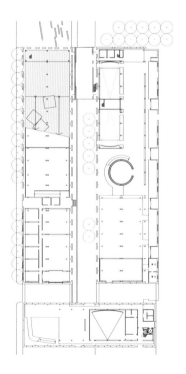
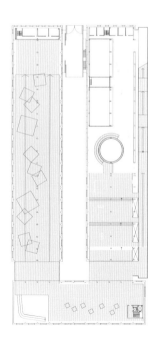

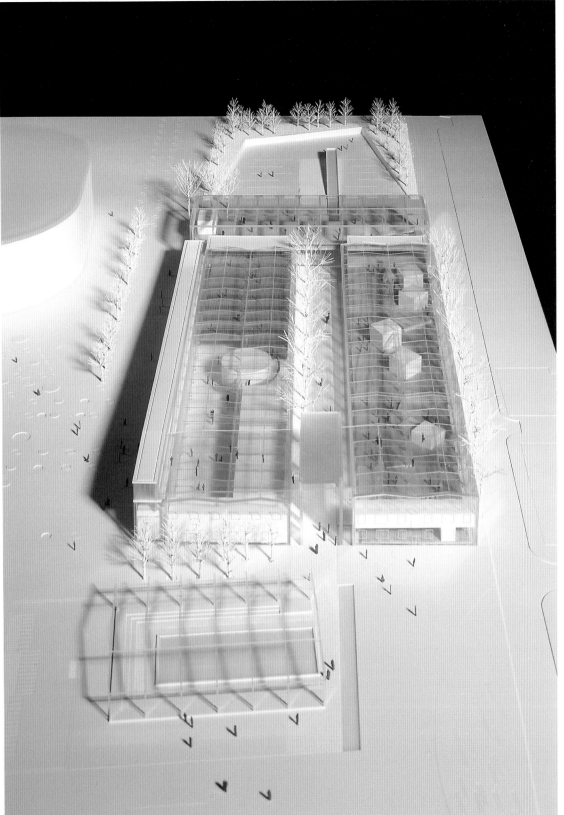

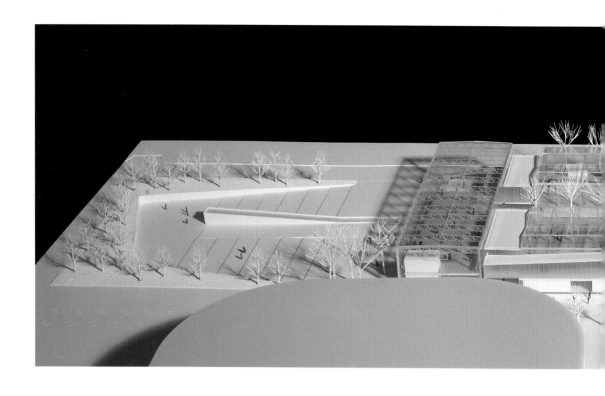
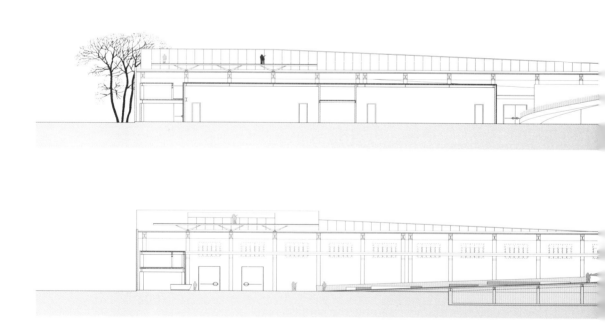

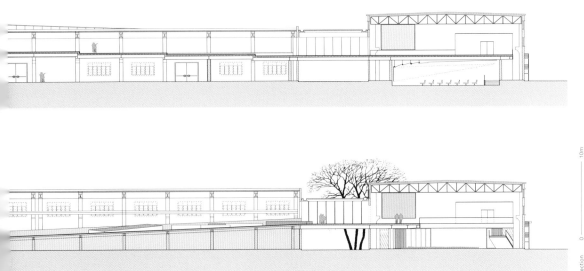

(top)model. (bottom)section

地域化

奥林匹克公园中心区下沉花园 [4.31]
北京图书大厦二期 [4.32]
西双版纳机场航站楼 [4.33]
巴塘人民小学宿舍 [4.34]
中国驻澳大利亚大使馆 [4.35]
中国驻印度大使馆 [4.36]
BIAD休息亭 [4.37]

Localization

Sunken Garden in the
Center Area of Olymipc Green [4.31]
The Second Phase
of Beijing Book Building Project [4.32]
Xishuangbanna Airport New Terminal [4.33]
Dormitory Design
of Batang School Campus [4.34]
Chinese Embassy in Australia [4.35]
Chinese Embassy in India [4.36]
BIAD Pavilion [4.37]

地域化：别处的建筑 / 刘宇光

建筑是人类文明的载体，用物质化的方式表达人对世界的认识和看法，实现过程就必须要和社会条件、生活环境及生产方式相结合。建筑不像汽车、飞机、iPad一样是全球通用设计，地域性差异往往是它得以产生的最直接的源泉。

UFo工作室长期承担着中国外交部驻海外使领馆的设计项目，工作地点从接近北极圈的冰岛到接近南极圈的澳大利亚，从俄罗斯的圣彼得堡到南亚次大陆的印度。虽然是同一类型的使领馆建筑，但处在不同的位置和文化背景下，就显现出差异化的取向和结果。使领馆建筑承载的一项重要使命就是使中国文明与驻在国文明相结合，建筑本身即是交流的使者。

中国驻澳大利亚大使馆项目开始于2005年，它位于澳洲首都堪培拉市首都区，在格里芬湖畔的城市公园里。堪培拉的总体规划是格里芬按照花园城市的理念完成的，经过100年的建设，已经成为一座名副其实的花园城市。首都区的规划十分严格，对基地内建筑和环境品质有很高要求，尤其对基地内的树木有明确的保护要求。我们的设计出发点是将中国的园林建筑和当地的花园城市体系相结合，用黑白灰的水墨化方式，让建筑融于自然，使景观成为主角；对于建筑本身，从材料和细节上反映当代中国建筑业的进步成果。为提高建筑的精度和可建造性，采用预制装配体系进行设计和建造，所有构造都在中国的工厂内预制，海运到现场安装，创造了建筑通用性生产和地域性特征相结合的模式。

Architecture is the carrier of human civilization, built by people to express their understandings and views of the world in a material way. Therefore, the implementation process of architecture must be combined with social conditions, living environment, mode of production and so on. Unlike cars, planes, or iPads which are products of universal design, normally it is the regional differences that inspire architecture.

UFo has for a long time been undertaking design projects for the Chinese overseas embassies and consulates, ranging from Iceland near the Arctic Circle to Australia close to the Antarctic Circle, from St. Petersburg in Russia to India on the South Asian Subcontinent. Although they are of the same architectural type, orientations and results vary with locations and cultural backgrounds. One of the important missions for overseas embassies or consulates is to promote the exchange between Chinese culture and other cultures, the building itself will be a symbol for such a mission.

The Chinese Embassy in Australia is located in the capital district of Canberra, the Australian capital, in the urban park by Lake Burley Griffin. This project began in 2005. Canberra's overall plan was developed by Griffin based on the "garden city" idea, which made Canberra, after 100 years of construction, become a veritable garden city. The planning within the capital area is very strict , imposing rigid requirements for the qualities of buildings and environment, particularly for protection of the trees. The starting point of our design is to combine Chinese garden style architecture with local system of garden cities, using black, white and grey colors to embody a Chinese ink painting, to have the building fit into nature, and let the landscape be the leading role. For the building itself, we focus much on materials and details to show the latest achievements in Chinese architectural industry. To improve architectural accuracy and constructability, a system for assembling prefabricated panels is adopted for design and construction - all structures were prefabricated in domestic factories, then shipped to install on site.

In the project of Chinese Embassy in Delhi, Indian capital, we expressed our thoughts on architectural regionalism through cultural dialogues. We used Chinese traditional patterns as elements to form the modular components, a cultural carrier of communication between Chinese and Indian civilizations. The building was designed white, so as to keep harmony with the color of local buildings, also to show our respects

在中国驻印度大使馆项目中,我们通过文化层面的对话,表达了对建筑地域性的思考。以中国传统纹样为元素,形成模块单元组件,作为中印文明交流的载体。白色的建筑既与当地建筑的颜色保持和谐,又是对在印度留下杰出作品的柯布西耶的致敬。为了适应南亚次大陆的气候,建筑采用过渡空间,营造层层递进的开敞式布局,让光、空气和水等自然元素渗入建筑,生出一种轻松愉悦的禅意。

除了海外项目,UFo在国内许多地域性特征明显的地区也有建筑实践活动,比如云南的西双版纳、内蒙古的鄂尔多斯,等等。2012年开始设计的巴塘县人民小学宿舍楼,位于四川甘孜州藏族自治区。这是一个公益集群设计活动,项目的地域条件和建设条件都和我们平常生活的环境有很大的差异。我们的设计采用多中心的庭院布局,为孩子们提供多种多样的活动空间,成为人介入环境的一种方式。由于所处地区偏远,建造难度和成本均高于平常地区,因此建筑着重在空间上最有效地利用资源,创造最大化的活动空间。场所的尺度、特征和布局方式也力图与当地生活习惯及自然条件相对应,避免标签式的地域主义符号。

UFo所进行的地域性建筑设计实践,始终遵循着普适性原则和特殊地域要求相结合的方式,而不是采用标签式的简单拼贴方法。我们希望透过建筑本体,反映我们对该地域的认知。

to Le Corbusier, a master who left outstanding works in India. In order to accommodate local climate, we used transitional spaces to create a progressive, open layout, which invites light, air and water to infiltrate the building, creating a zen-like relaxation.

In addition to overseas projects, we have also carried out practices in many domestic areas with obvious regional characteristics, such as Xishuangbanna in Yunnan, Erdos in Inner Mongolia. The project in Ganzi Tibetan Autonomous Region of Sichuan, People's Boarding School Dormitory Design of Batang County, began in 2012. This is a non-profit group design program, the regional and constructional conditions are very different from what we are familiar with. Eventually we proposed a multi-center courtyard which allows a wide variety of activity spaces for children, an approach for people to get into the environment. As it is a remote area, construction difficulty and cost are both higher than usual, the construction effect is mainly reflected on the space, involving the most efficient use of resources to create maximized activity space. Meanwhile, the scale, characteristics and layout of the site are designed to accord with local habits and natural conditions, avoiding a label-style regionalist symbol.

Our regional architectural practices have always been combining the universal principle with specific regional requirements, rather than a simple label-style collage. We hope to reflect our understanding of the region from the building itself.

奥林匹克公园中心区
下沉花园

Sunken Garden in the Center Area of Olympic Green

4.31

用地面积:
45 000平方米
建筑面积:
6 394平方米
建筑高度:
8.495米
设计时间:
2006—2007
建造时间:
2007—2008

该项目位于北京奥林匹克花园中心区内,共有七个下沉式院落,设计要求体现中国元素。共有五家设计单位参与了这个项目,每家从不同的角度对中国传统文化进行了深入的诠释。UFo工作室负责总体规划与技术协调,以及一号、四号和五号院设计。紫禁城和四合院是北京城的代表。在以往的等级社会中,它们被高耸的红墙截然分开。今天,随着多元、开放、平等、和谐时代的到来,红墙的禁止功能被交流功能所取代,这条难以逾越的边界开放了,宫城禁地和民间胡同相互融合,在奥林匹克公园里诞生出一个新的地域景观——开放的紫禁城。中国传统建筑的每一个单位,基本上是一组或多组围绕一个中心空间(院子)而组织成的建筑群。我们从围墙入手,将一个单个围合的院落纵向切开,再首尾相接,使闭合型变成开口型,再不断相接,将闭合的"禁城"转化为开放的"非禁城",最终产生了无限开放但保持原始空间尺度和感觉的意象。开放的紫禁城既保留了北京原有的意象,又通过红墙、灰墙重构了全新的动态空间,使人能从这个新的场所中体验中国的传统文化。(下图: 从封闭到开放的空间概念)

Site Area:
45,000m²
GFA:
6,394m²
Building Height:
8.495m
Design Time:
2006-2007
Construction Time:
2007-2008

Situated in the central area of Beijing Olympic Garden, the project covers 7 sunken courtyards whose design elements interprete the representation of Chinese elements. Totally, 5 companies participated in this project and each of them made a profound expression from different views on the traditional Chinese culture. UFo took the responsibility on the overall planning and technical coordination, as well as the design of No.1, No.4 and No.5 sunken gardens. The Forbidden City and the quadrangle courtyard is the representative of Beijing. However, with the time of pluralism, openness, equality and harmony comes the wall's function banning is replaced by its exchange function, so the boundary which is hard to override is open. The imperial palace merges the civil alley thus the latest regional landscape debuts in the garden - the opened "Forbidden City" is unveiled. Each unit in traditional Chinese architecture is basically an architectural complex in which one group of units or multiple groups of units encompass a center space (a courtyard). Beginning with the enclosing walls, we vertically cut through a singularly enclosed courtyard, then re-join it from end to end, which thus changes from a closed type to an open one, and then continue to join it from end to end, which thus transforms an enclosed "forbidden city" to an open "non-forbidden city," resulting in such an image as is infinitely open while keeps the original spatial scale and sense. An open forbidden city not only maintains Beijing's original image, but also re-structures a whole new dynamic space through both red and grey walls, which enables views to experience traditional Chinese culture through such a new site.
(Above: spatial concept from close to open)

七个近似的院落，由南至北形成层次递进的空间，如同北京传统住宅院落的纵向发展结构，体现了中国传统文化的意象。它们从不同的角度诠释了中国传统文化一号院御道宫门，表现了城市开门的宏大场景；二号院古木花厅，拉近了人的尺度，让人体验地方民居文化；三号院礼乐重门，使人从礼乐活动中感受中国古老的文明；四、五号院穿越瀛州，在穿越隧道的前后过程中体会绿色瀛州；六号院合院谐趣，展现了四合院作为公共活动空间的热闹场景；七号院水印长天，刻画了皇家园林中的传统运动场面。

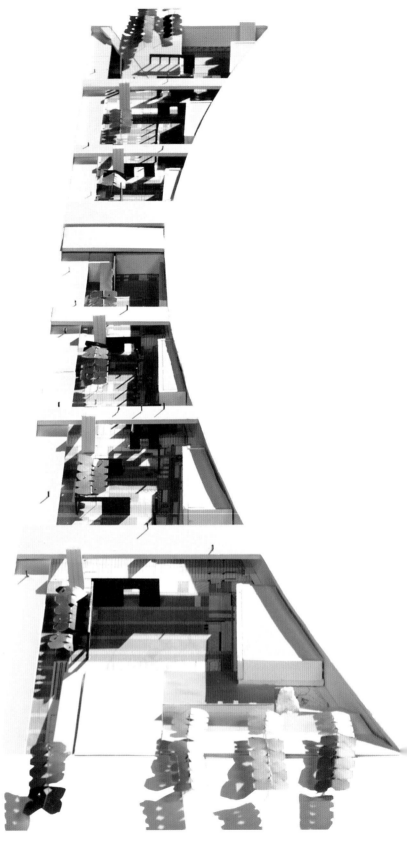

The 7 homogeneous courtyards form the progressively layered space featuring the longitudinal growth structure similar to the traditional house courtyards in Beijing. This embodies the special intention of traditional Chinese culture and even illustrates this culture on different views: The "Yudaogongmen" of No.1 yard presents the grandness of city's opening its gate; The "Gumuhuating" of No.2 yard is close to human scale who enjoy experience the civic-house culture; The "Liyuechongmen" of No.3 yard makes the people taste the ancient Chinese civilization during the imperial celebration; The "Chuanyueyingzhou" of No.4 and No.5 yards makes the travelers go round the green "island" through a special tunnel; The "Heyuanxiequ" of No.6 yard represents a live sightseeing of quadrangle courtyard; And the water and sky of No.6 yard depict the traditional sports scene in the royal garden.

(left) concept model; (top-right) main gate of courtyard No.1; (bottom-right) night view of courtyard No.1

(左页图) 概念模型；(右页上图) 一号院大门；(右页下图) 一号院夜景

一号院以故宫午门前广场为设计意象，所包围的空间是一个举行仪式的露天礼堂，正对大台阶入口是午门意象的宫门，结构形式采用钢结构梁柱体系。门架高11米，宽36米。门扇南面户外全彩LED显示屏，可播放18米宽5米高的巨幅画面，门架顶端南侧为挑檐深远的棚架，143根铝型材一端与门架铰接固定，型材交错布置，相互联接，形成两条曲线，抽象表达了反宇向阳的中国建筑神韵。

Hall 1 in the Imperial Palace adopts the front square at the Meridian Gate as its image of design. The space which it encompasses is an open-air auditorium which serves to perform ceremonies. Faced with the entry of the grand staircases is the Palace Gate of the Meridian Gate image, whose structural style adopts the steel-work and beam-column system. The gantry mounting is 11 meters in height, and 36 meters in width. The outdoor full-color LED display screen at the south of the door leaf can display pictures which are 5 meters in height and 18 meters in width. The south side of the top of the gantry mounting is a canopy frame which has far-reaching cornices. One end of each of the 143 pieces of aluminum profile is hinged and fixed on the gantry mounting. The materials are in staggered arrangement, interconnected with one another, in order to form two curves, which abstractly shows the Chinese architectural charm that the concaved roof upward is mutually mingled with the heaven above, as yin and yang in nature, to blend into a harmonious whole, making those who live inside the building fused into heaven and earth.

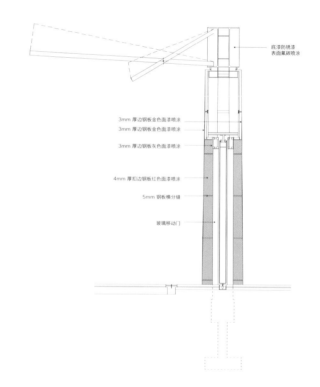

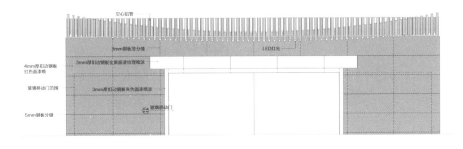

(top-left) tradition Chinese elements in courtyard No.1; (top-right) view from courtyard No.1 to water cube; (middle-right) plan of red wall; (bottom-left) section of red wall; (bottom-right) elevation of red wall

北京图书大厦二期

The Second Phase of Beijing Book Building Project

4.32

用地面积:
4 833.22平方米
建筑面积:
43 122平方米
容积率:
3.6
建筑高度:
38米
设计时间:
2005—2006
建造时间:
2007

本项目位于北京城市中心的西单地区,南邻西长安街,西邻西单的城市广场。图书大厦一期工程建成于上世纪90年代。根据业主的计划,扩建内容包括在一期的北侧扩建一幢新楼,同时对现有建筑进行全面整合,使其与扩建部分共同形成一个面向未来的城市新地标。面对这样一个设计任务,我们希望在方案中体现以下几个原则:改扩建后的建筑首先应该与城市保持和谐的关系,与城市保持对话,并具有足够的城市亲和力;新的建筑形象需考虑城市发展的连续性,通过适当保留原有建筑的历史痕迹,使城市保持应有的记忆;充分考虑原有建筑的条件和预算限制,同时体现可持续发展建筑设计理念。针对现有建筑与扩建部分的整合,设计上摒弃全面包装、改头换面的办法,而采用三维立体的方式,通过镂空的建筑表皮处理,根据不同部位的功能条件,对建筑四个立面和屋顶进行整合,形成一个全新的同时又适当保留原有建筑痕迹的建筑新形象。(下图: 设计概念分析图)

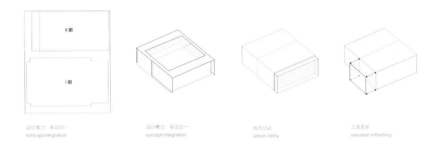

concept integration　　concept integration　　urban lobby　　elevation refreshing

Site Area:
4,833.22m²
GFA:
43,122m²
FAR:
3.6
Building Height:
38m
Design Time:
2005-2006
Construction Time:
2007

The project is located in Xidan which is city center of Beijing. It neighbors West Chang'an in south and Xidan Urban Plaza in west. The phase 1 of Book Building was built in 1990s. Depending on the owner's planning, its expansion includes building a new tower on the north of this term; at the same time, the constructor will fully integrate the existing buildings to create a future oriented new urban landmark incorporating the new parts. In this case, we wished to reflect the following principles in the solution: Both the new and renovated buildings ought to make a harmonious relationship with the city, so as to respond the city and have an enough urban affinity; The latest building façade needs take account of the city's growth continuity; so as to make the required urban memory remained with properly keeping the historic sites of old buildings; Fully considering the states and budget limits of existing buildings, in order to show the design concept of sustainability in architecture. On the view of the integration of the new and renovated buildings, Ufo did not changing the facades design completely, but incorporate 4 facades with the rooftop depending on different parts' conditions. In details, a three-dimension processing is taken with the treatment of hollowed-out building skins, with an aim to create the new appearance of old buildings whose historic tastes are remained. (Above: design concept analysis)

建筑表皮的研究，通过单一图案的叠合和细部构造处理，获得复杂、丰富的肌理效果。建筑外墙系统包含三个层次：内立面、外层表皮和中间的结构。

UFo had conducted a research on the building skin as follows: With the overlapping of single patterns and the detailing of structures, UFo won the complicated but rich textiles. The system of envelopes on building covers 3 layers: the indoor façade, the outside skin and the middle structure.

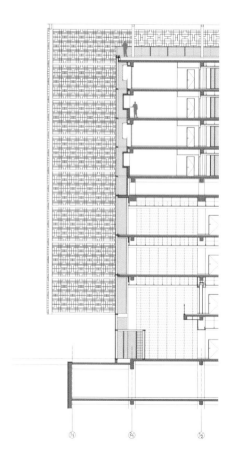

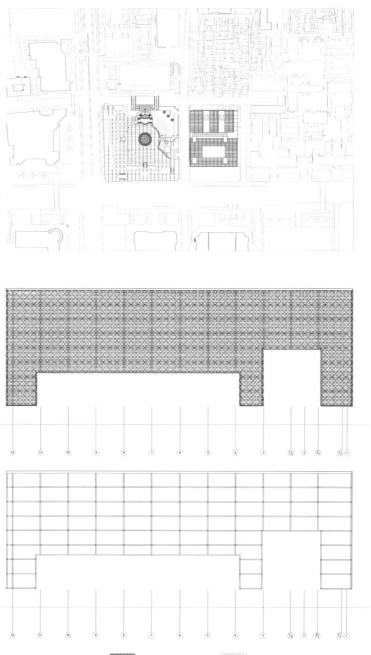
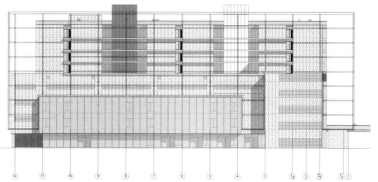

(top-left)detailed structure; (top-right)site plan; (bottom-right)north elevation

西双版纳机场航站楼

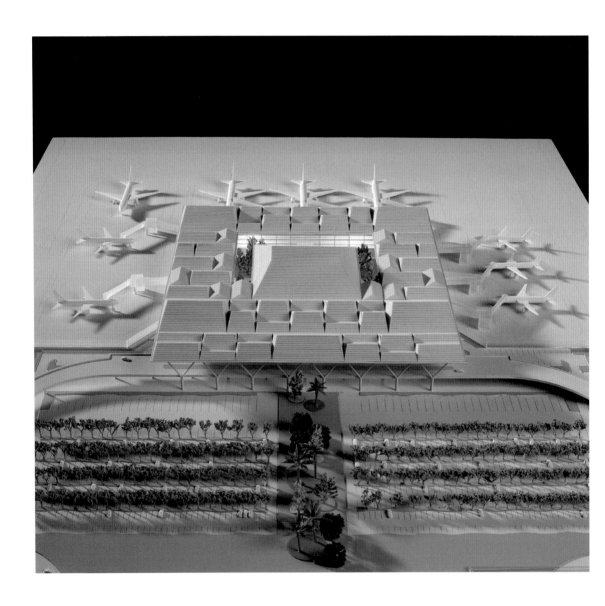

Xishuangbanna Airport New Terminal

4.33

用地面积:
15 500平方米
建筑面积:
33 200平方米
容积率:
0.32
建筑高度:
34米
设计时间:
2008
建造时间:
2008

西双版纳机场位于景洪城区西南。新航站楼和站前广场位于现有机场航站区北部的新征用地上，占地102 500平方米。该用地是政府权衡多种因素后，拆迁传统村寨而得到的建设用地。新航站楼构型采用了独创的"口"字形，东部布置陆侧主要功能空间，西、北、南三面围合形成的"C"形布置空侧主要功能房间，中心为地域特色共享大厅，室外花园呈"C"形环绕共享大厅。"口"字形构型充分利用了空侧站坪进深，这一变形指廊式构型结合了指廊式航站楼与卫星式航站楼的优点，不但使内部的流线顺畅新颖，飞机运行效率提高，而且在不占用过多面宽的条件下，三面泊机，近机位多且候机距离较均衡。航站楼体形和立面设计灵感来源于场地中原有的典型民居——曼南村寨。以"傣寨"为设计主题，是把当代设计和地域文化综合的一次尝试。设计过程中对屋顶的形式系统作了大量研究，内容包括外部形式的秩序和变化，外部形态和内部空间的关系，屋顶形式系统单元的基本尺寸、采光和排水等。

(下图: 设计概念分析图)

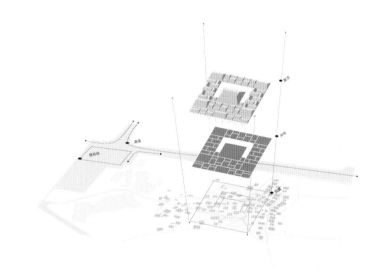

Site Area:
15,500m²
GFA:
33,200m²
FAR:
0.32
Building Height:
34m
Design Time:
2008
Construction Time:
2008

Xishuangbanna Airport is in the southwest of Jinghong City. The new terminal building and the airport square are at a newly certified site in the existing Xishuangbanna Airport, with an area of 102,500 sqm. The site is a construction land that which is traditional village demdition and relocation with considerations. In structure design, that new terminal building applies a unique "square" pattern, whose east side is arranged with the main function features, and rest 3 sides make a "C" loop of void space for function room. In addition, the center is a characteristic lobby and the external garden surrounds this lobby in the "C" shape. The "square" pattern makes full use of the void space. After the innovation, the corridor structure is combined with the advantages of a corridor terminal and a satellite-style one, making the internal flow smooth and novel and raising the flight operation efficiency. Moreover, it makes the parking on three sides without more sides being occupied. That is, the 3 sides are used to land the flights with more parking stands and more balanced spacing. The inspiration of shape and façade designs, originates in a typical vernacular dwelling, the Mannan village. UFo set "Dai village" as the design theme attempting to integrate contemporary designs into the culture. And for the form of terminal's roof, UFo had studied the design routine and address the order and change of external form, the relationship between the indoors and outdoors, as well as the basic size, lighting and drainage of the roof pattern system.

(Above: design concept analysis)

场地中原有的曼南村寨是典型的当地民居，因为新航站楼的建设而被拆除。新航站楼的屋顶概念来源于曼南村寨，希望此类形象能够给到港的游客一个直观的地域印象，给当地人留住一份历史记忆。新航站楼的屋顶形似村寨中的单独屋顶组合在一起时形成的当代的、整体的形象。同时也能和老航站楼形成良好的呼应关系。

值机大厅标高为八米，候机区为七米，旅客的主要流线（含远机位候机）都是从高处往低处走，舒适性强，更为人性化。从共享大厅穿过透明连桥时，有置身花树丛中的感觉。

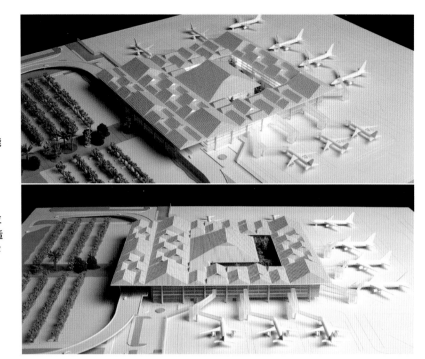

The existing Mannan village is a typical vernacular dwelling. It was moved due to the construction of a new terminal, the roof solution of which originates in this village with an aim to make the travelers have a direct impression on the region and the locals a historic memory. The new terminal's roof presents a contemporary and complete image that seems to make the single roofs of village into combination. This also echoes well with the old terminal.

The check-in lobby is 8m hign and the waiting area is 7m high both in elevation, and the circulation of passengers (apron) start from the higher area to the lower, with the comfortable feeling and the personality - Passing through the lobby to the bridge made from transparent materials, the people have a feeling of staying in flowers.

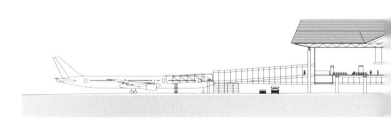

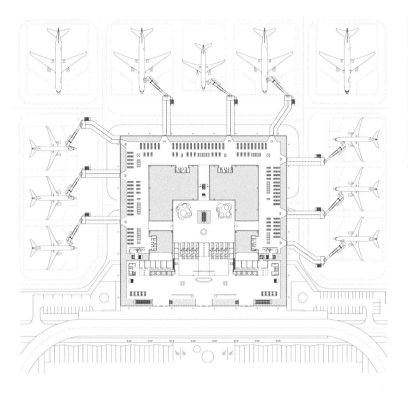
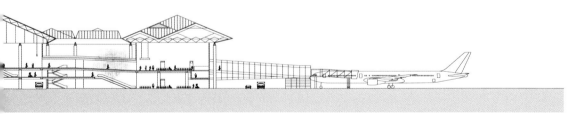
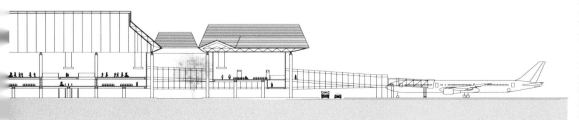

(top-left)model; (top-right) departure level plan; (bottom)section

巴塘人民小学宿舍

Dormitory Design of Batang School Campus

4.34

用地面积:
6 377.6 平方米
建筑面积:
6 500 平方米
容积率:
1.02
建筑高度:
13.7 米
设计时间:
2012—2013
建造时间:
2013—

巴塘教育园人民小学宿舍位于教育园区内人民小学的东南角,东侧为教学主楼,北侧为食堂和运动场。宿舍采用小尺度的组群式布局,形成多层次的活动空间,为具有百年历史的巴塘人民小学提供了新的活力。宿舍按东西分为男女生两个部分,中央为公共活动广场,东南西北设置了四个院落空间,突出巴塘民居的自然特征,改变了宿舍建筑常见的呆板体量。宿舍共有160间,以3~4层建筑为主,采用一侧外廊、一侧阳台的双面采光模式,提供健康自然的生活环境。建筑墙面上的不同颜色,既增强了楼体的识别性,又体现出藏式民居的特色。(下图: 总平面图)

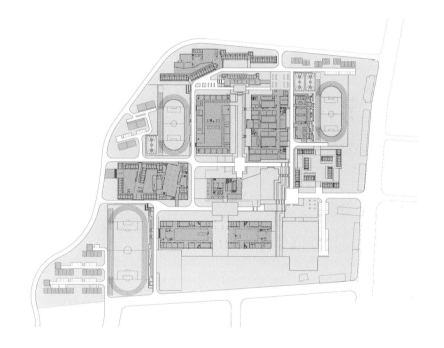

Site Area:
6,377.6m²
GFA:
6,500m²
FAR:
1.02
Building Height:
13.7m
Design Time:
2012-2013
Construction Time:
2013-

The People's Preliminary School in Batang Education Garden places its dormitories in the southeast corner of campus. The east of dormitories is a main teaching building, and the north covers the dining hall and the playground. This hall sees a small-scale cluster layout, by which a multi-layer activity space is formed for the school of 100 years filled with vigor. The dormitories are divided into two parts for male and female. The center is a public square 4 corners of which are set with 4 yards. This stresses on the natural features of vernacular dwellings in Batang and changes the regular volume of usual dormitories. In order to give a green environment, the new dormitories have 160 rooms which are focused on the buildings of 3-4 levels with an outside corridor and a balcony for the contrary sides. The façade is painted in different colors to make the identity and the Tibetan-dowelling characteristics. (Above: site plan)

以"旋转"为空间造型机制，形成错落有致的空间。每组建筑长边按"L"形被折成两部分，因此建筑尺度较小而内部空间较多，形成中间广场、入口街道及内院空间三个层次，并且可以循环往复。实体和空间互为因果，虚实相间，形成多画面的活动场景。

Taking the "rotation" as a spatial-pattern system, the dormitories are formed into the overlapped and coordinated spaces. Each a group of buildings' long sides is divided into two parts in the "L" shape. Therefore, the buildings feature a small size and more spaces, including the central square, the street and the in-yard space all of which are able to be repeated. The entities and the spaces interact as both cause and effect, with the combination of abstraction and reality, to form the multiply images of activities.

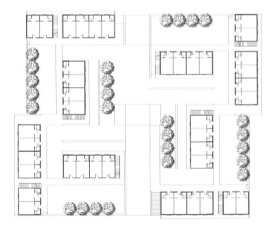

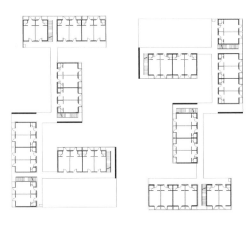

(top-left) during construction; (top-right) construction site; (bottom, from left to right)1F, 2F, 4F plan

（左页上图）施工过程；（右页上图）建设用地；（下图，由左至右）一、二、四层平面图

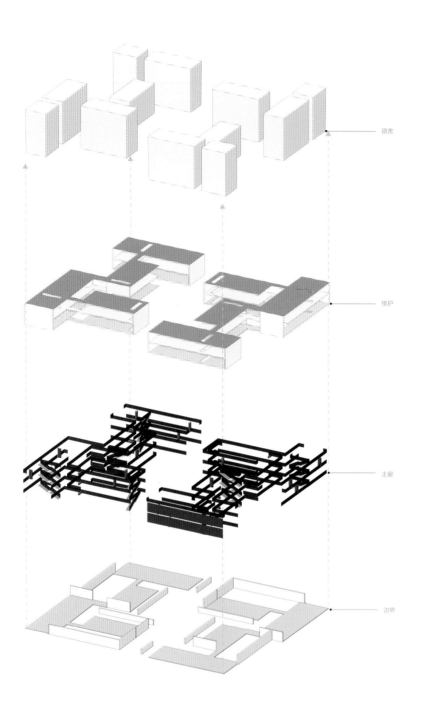

(左页图)体块分析;(右页图)效果图

(left) block analysis; (right) rendering

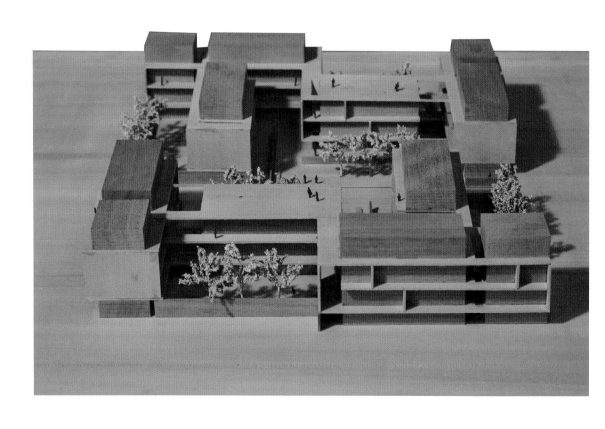

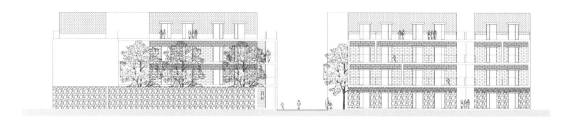

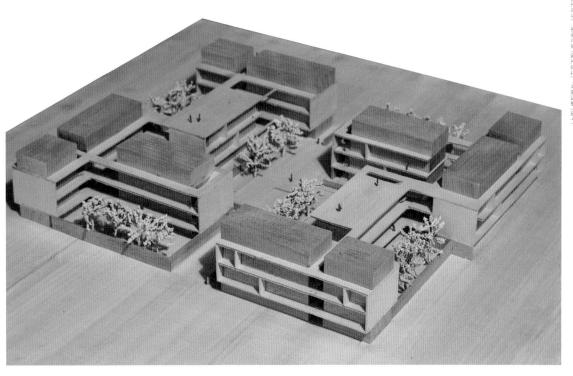

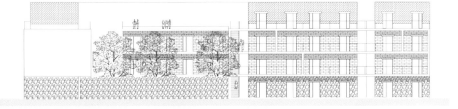

(top) model; (bottom-left) south elevation; (bottom-right) east elevation　0———10m

中国驻澳大利亚大使馆

Chinese Embassy In Australia

4.35

用地面积:
18 000平方米
建筑面积:
6 858平方米
容积率:
0.3
建筑高度:
12米
设计时间:
2006
建造时间:
2011—2013

中华人民共和国驻澳大利亚大使馆新馆位于首都堪培拉的亚拉鲁姆大区128分区地物地段地皮。基地位于现有使馆馆舍西侧,两馆中间为Flynn Drive 公路及城市绿化带。新馆的设计尊重并借鉴周边环境和"使团区"的特性,致力从尺度到材质上与相邻建筑融为一体。同时,整体建筑设计和规划遵循了中国传统建筑规划的原理,并加之以现代诠释。建筑物的布局、室内外空间秩序条理以及园林的运用从整体上达成了建筑与园林的和谐。新馆建筑形体传达了中国传统建筑形式,同时又满足现代生活和工作的需要。建筑表面处理简洁,旨在为大使馆工作人员提供安全、合理的办公空间。室内空间开敞,室外景致一目了然。开放式和封闭式的办公室环绕室内庭院。园林设计是大使馆新办公楼建筑设计的重要组成部分。三个建筑环绕主花园,既有宁静感,又富安全感。通过中西园林设计理念的运用,自然、含蓄的园林将与使馆区的总体自然环境相得益彰。(下图: 模型照片)

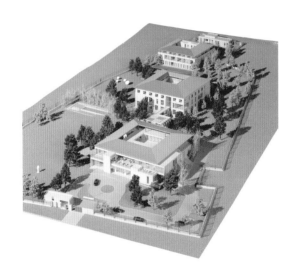

Site Area:
18,000m²
GFA:
6,858m²
FAR:
0.3
Building Height:
12m
Design Time:
2006
Construction Time:
2011-2013

The new embassy of the People's Republic of China in Austria is located in No.128 Area of Yarralumla, Canberra. The site is in the west of existing embassy, with a middle zone of Flynn Drive Road and the green belt. The design of new embassy treasures the surrounding area and the "mission area" characteristic, focused on the integration of size, material and neighbor. Meanwhile, the whole design and planning follow and modernly interpret the principles of traditional Chinese building plans. The layout, inside and outside spaces orders and gardens are worked out to bring out the harmonious relationship between building and garden. The pattern of new embassy reflects that of a traditional Chinese building and satisfies the needs of modern living and working. The façade is tidy on treatment to provide the staff with safe and reasonable office spaces. The indoor space is open for an easy external landscape. The opened and closed offices surround the indoor yard. The garden is designed as one of the key building parts. There are 3 buildings around the main garden, where is peaceful and quiet. With the use of Chinese and western landscapes' design concepts, the natural and implicit garden will compliment the overall natural environment. (Above: model)

澳大利亚首都堪培拉由芝加哥建筑师格里芬规划完成,是花园城市的代表作。使馆基地紧邻城市公园,基地内有很多树木需要保留。因此布局上采用见缝插针的办法,尽量不破坏现有环境,也与花园城市的总体风貌保持一致。

Canberra, the capital city of Australia, was planned by Griffin, architect from Chicago. It is a representative work of city garden. The new embassy occupies an area neighboring one of the city's parks. The side is required to preserve lots of trees, so the layout applies a method of cutting corners in its enthusiasm, trying its best not to destroy the existing environment and make the embassy consistent with the garden-style city.

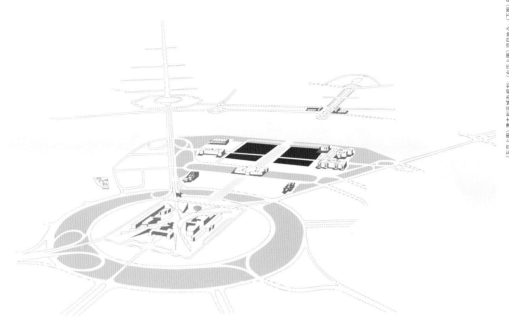

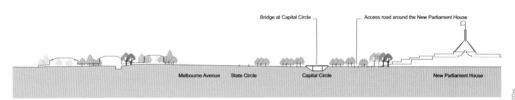

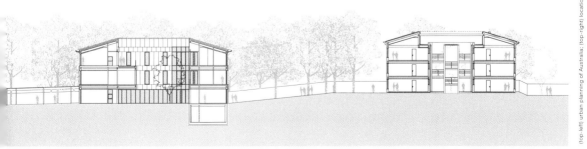

中国驻印度大使馆

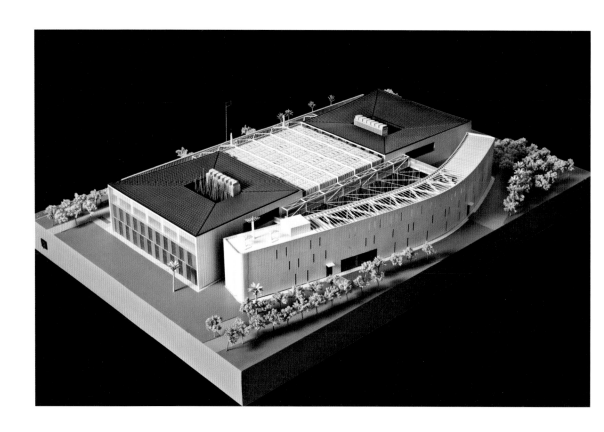

Chinese Embassy In India

4.36

建筑面积:
13 500平方米
建筑高度:
15米
设计时间:
2011—2014
建造时间:
2013—

中国驻印度大使馆是中国驻外使馆中等级较高的单位,外交部提出了建筑设计要传承东方文化意蕴,展示国家风采,传统和现代相结合等要求。建筑主要包含办公、外事接待、后勤服务三部分功能。建筑通过群组的方式形成整体,既与环境融合,也带来建筑空间的仪式感和庄重气氛。这种通过组合方式产生的空间威严,不像单纯通过增加建筑体量的方式那样给人以压迫感,我们试图保留中国传统建筑中特有的灵动性。设计中充分考虑了当地气候的影响,倡导绿色理念,并按照绿色三星标准实际操作:通过增加屋面气流交换层降低屋面温度;通过拔风中庭改善室内气流循环;通过建筑立面的凹廊、遮阳百叶以及建筑入口空间的遮阳棚架,对极热地区的气候进行回应。这座建筑也将成为中国驻外建筑中第一个绿色三星建筑。建筑强调以人为本,试图打破中国当代"官式建筑"的范式,通过人性化的设计,强调空间的公共性、流动性和亲和力,为外交官的生活注入新的活力。(下图:总平面图)

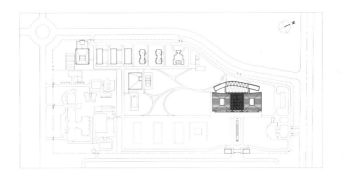

Building Area:
13,500m²
Building Height:
15m
Design Time:
2011-2014
Construction Time:
2013-

The Chinese Embassy in India ranks high among the Chinese embassies in foreign countries. In this case, the Ministry of Foreign Affairs puts forward the requirement of inheriting the cultural implication, showing the national styles and combining the tradition and modernity by means of building designs. The architecture contains three major functions: office, foreign affairs and logistics. It is integrated with cluster buildings as well as the environment, bringing about a sublime atmosphere. This will not make people feel oppressed as the way of simply enlarging the mass. The designer UFo attempted to reserve the unique flexibility as a traditional way, taking account of the impact of local climate, advocating the green ideas and carrying out operations according to the 3-star green standards: Lowering the roof temperature by adding the exchanging layers of roof airflow; Improve the circulation of indoor airflow by ventilating the yard; and Respond to the extremely hot climate with concave corridors and shade louvers in the façade, ad shade scaffolding in the entrance space. This building will be the first 3-star green architecture among the Chinese embassies. It plays an emphasis on the people's health and tries to break the pattern of Chinese modern official architecture. Through the humanized design, the architecture is focused on the public nature, mobility and affinity of the space and injects new energy into the diplomats. (Above: site plan)

通过坡屋、合院、四水归堂以及立面的"侧角升起之制"等传统建筑语言，营造建筑的历史认同感。

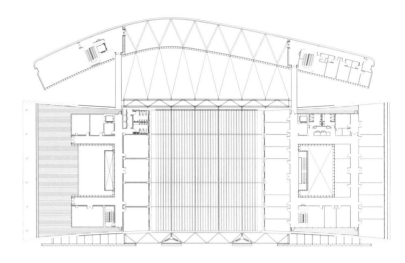

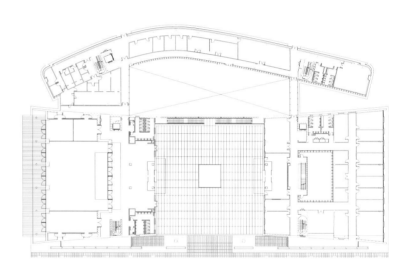

The architecture creates historical identity with its traditional architectural language, including the flat roof, the courtyard house, circulating water system in the courtyard and the manufacture of risen side angles in the facade.

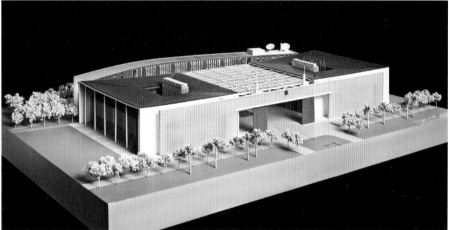

(top-left) 3F plan; (bottom-left) ground floor plan; (right) model photos 0 —— 10m

建筑设计还涵盖了纹样设计。我们没有直接使用中国传统上的某个纹样,而是通过变异、抽象,衍生出符合现代审美的新纹样。

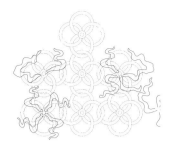

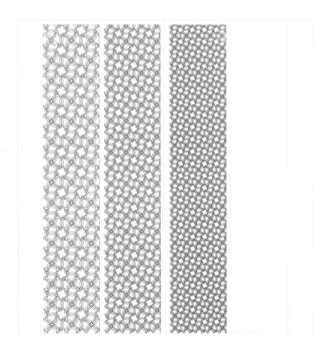

The architectural design also contains the pattern design. We did not directly pick some particular patterns from the tradition, but rather generated new ones through variation and abstraction, coming up with some that fit modern aesthetics.

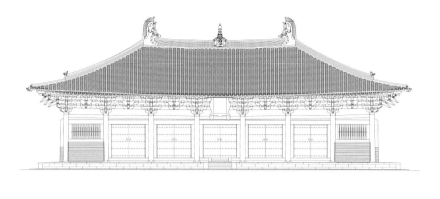

轴001-027	
轴028-054	
轴055-081	
轴082-108	
轴109-131	
轴132-158	
轴159-184	

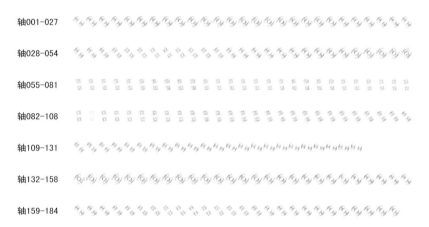

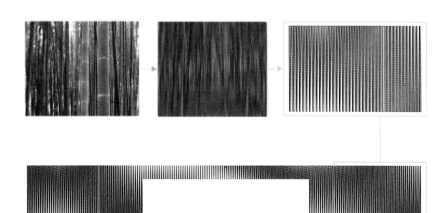

(top-left) carving patterns of Tian'anmen; (bottom-left) pattern Evolution; (top-right) final design rendering; (middle-right) analysis of elevation ascending; (bottom-right) parametric study on east elevation texture

(左页上图) 天安门山花纹样;(左页中图) 纹路演变;(左页下图) 最终设计效果;(右页上图) 立面升起分析;(右页下图) 东立面肌理参数化研究

BIAD休息亭

BIAD Pavilion

4.37

用地面积:
35平方米
建筑高度:
3米
设计时间:
2006
建造时间:
2006

这是一个在既有环境中增建的亭子,是北京市建筑设计研究院为员工提供的一个休息交流的半室外空间。我们希望它既能与环境相融合,又有自身的完整性。我们从现有的树木、铺装、座椅、绿篱中提取了基本网格,建立了一个连续的构成体系。它的支撑体系是连续的,根据空间需要选择用透明与半透明,既有连续的整体感,又有单元的灵活性。人们在这个亭子当中体验一年四季的变化。(下图: 设计概念分析图)

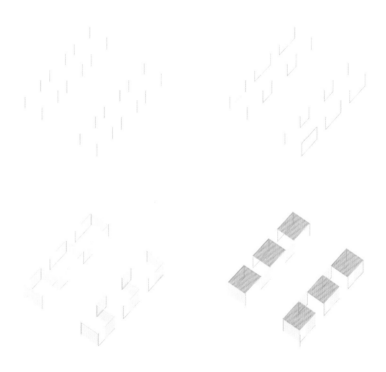

Site Area:
35m²
Building Height:
3m
Design Time:
2006
Construction Time:
2006

This is a pavilion built in an existing environment, a semi-outdoor space provided for the employees to relax and have a chat. UFo expected that it will not only fit into the environment, but also have own integrity. A basic pattern was extracted from the existing trees, pavement, seats and green fence to build a continuous system. Its supporting system is consecutive with needing to choose the transparent or semi-transparent materials, which are of continuously whole taste and of a unit's flexibility. People will experience in the changes of four seasons here.
(Above: design concept analysis)

从既有的广场铺装、树阵和室外座椅当中,寻找整体性逻辑,从而把独立的景观元素融合起来,形成了一个独特的公共活动场所。均质化空间,可以保持私密性的独立空间属性,也便于群体的交流。钢结构的搭建、玻璃尺寸的选择、排水方式的确定,都回应了材料和场地的最基本诉求。

The pavilion attempts to seek the overall logic from the existing square's pavement. Trees and outside seats, so as to form a unique public area. The homogenized space can keep the independent spatial attribute of a privacy, and become flexible in the exchange. The built steel structure, the selected glaze size and the confirmed drainage together make response to the basic appeals of material and site.

同行者

Cooperators

OMA建筑事务所 [5.1]
盖里事务所 [5.2]
扎哈·哈迪德建筑师事务所 [5.3]
KPF建筑师事务所 [5.4]
华汇设计（北京） [5.5]
SAKO建筑设计公社 [5.6]
KCAP建筑与规划事务所 [5.7]
景观都市主义工作室 [5.8]
BIAD第四建筑设计院结构设计团队 [5.9]
Speirs & Major事务所 [5.10]
克里斯蒂安·克雷兹建筑事务所 [5.11]

Office for Metropolitan Architecture [5.1]
Gehry Partners LLP [5.2]
Zaha Hadid Architects [5.3]
Kohn Pedersen Fox Associates [5.4]
HHD_FUN [5.5]
SAKO Architects [5.6]
KCAP Architects&Planners [5.7]
LAUR Studio [5.8]
Structure Design Team, BIAD Architectural Design Division No.4 [5.9]
Speirs and Major Associates [5.10]
Christian Kerez Zurich AG [5.11]

中国的建筑设计市场向世界开放以后，吸引了众多国际大师和著名事务所来华开展设计业务。UFo工作室和其中的代表人物，如库哈斯、扎哈·哈迪德、诺曼·福斯特、盖里等均有过合作，并且发展成为长期的合作伙伴；他们的职业精神和素质，以及对细节的严谨态度，对于提升中国建筑的整体水平有重要的作用。UFo在自己独立承接的项目中，也逐步采用国际合作的方式，自主聘请城市设计、数字设计、灯光、景观、室内设计等国内外优秀的团队为合作伙伴，走专业化合作的共赢之路。

本章，我们邀请UFo的12位同行者，为双方的合作撰写评论。

As Chinese architectural market has opened to the world, it has been attracting lots of international architects and well-known design firms. UFo studio has cooperated with many top architects such as Rem Koolhaas, Zaha Hadid, Norman Foster, and Frank Gehry et. With the long term relationships, UFo studio has been practicing and learning from their professional dedication and passion, meticulousness with the detail work which has been pushing the quality of the Chinese architecture into the next level. UFo also starts hiring international consultants on its own projects in urban design, digital design, lighting, landscape and interior design to achieve a win-win strategy in professional design cooperation.

In this chapter, we invite twelve of UFo's collaborators to write reviews from the previous cooperations.

OMA + UFo

OMA是一家走在世界前沿的国际合伙人事务所,专注建筑设计、城市规划及文化分析。创立30多年来,一直致力设计和建造大楼以及规划方案。事务所目前由七位合伙人领导,在鹿特丹、纽约、北京及香港均设有办公室,并将于多哈成立新办公室,员工总数近340名。

姚东梅,纽约注册建筑师,现任OMA北京公司总经理。

2004—2006年,OMA与BIAD UFo合作了北京图书大厦改扩建项目。此项目是OMA在2003年国际竞赛中的中标项目。方案的设计理念是,结合二期工程,对现有图书大厦进行立面及内部空间格局改造,使新旧建筑契合如一,内外焕然一新,让图书大厦成为长安街上的亮点,并注重与长安街风貌的协调。当时计划在2008年奥运会之前完工。由于地处长安街,此项目十分具有挑战性。BIAD UFo作为原北京图书大厦的设计方,此次以中方合作建筑师的身份,与OMA密切合作,旨在探索中外设计师合作的新的可能性。在方案设计期间,BIAD UFo的设计团队,多次赴OMA鹿特丹本部参加设计工作站会议,与OMA团队共同探索多种方案的可行性,同时为外方设计团队更好地理解中国的规范法规提供了有效的指导。当方案的审批工作遇到困难时,UFo的设计团队始终与OMA的设计团队站在一起,不懈坚持,积极协助与业主以及规划部门的沟通与报批审批工作。遗憾的是由于业主的变更,方案的推进受到影响,最终设计方案未能实施。纵观UFo十年来的作品,可以感受到这个工作室对建筑事业执着的追求,对服务品质和设计创新的不断超越,和对当代城市建设的贡献。我们衷心地期待着BIAD UFo在新的起点,承上启下,迈向新的辉煌。

北京图书大厦

5.1

Beijing books building

Yao Dongmei, registered architect in New York, general manager of OMA's Beijing office.
OMA cooperated with BIAD UFo on the Beijing Book Building project from 2004 to 2006. The Beijing Book Building enlargement & reconstruction project was won by OMA in the international competition in 2003. The design concept of the proposal of OMA is that combined with the second stage of the Book Building project, reconstruct the façade and the interior spatial pattern of the existing book building, integrate the old and new architecture, and enable it to take an entirely new look inside and outside, thus making the Book Building the highlight of Chang' an Avenue, take on a new look of the cultural architecture in Beijing, focusing on the coordination of styles on Chang' an Avenue. The plan then was to complete the project before the 2008 Olympic Games. This is a very challenging cultural project: located on Chang' an Avenue. As the designer of the original Beijing Book Building, BIAD UFo is the architect OMA is cooperating with, and BIAD UFo has close cooperation with OMA, aiming to explore the new possibility of cooperation between Chinese and foreign designers. During the proposal design, the BIAD UFo design team went to Rotterdam, the base of OMA, to take part in the design workstation conferences on several occasions, explored the feasibility of several proposals along with the team of OMA, and provided effective guidance for the foreign design team to better understand the specifications & regulations in China. When there were difficulties in the approval of the proposals, the UFo design team always stood together with the OMA design team, persistently and actively assisted the communication and approval with the owners and the planning department. Unfortunately, due to a change of the owner, the advancement of the proposal was affected, and the final design proposal could not be completed.Throughout a decade of works of UFo, we can feel the studio' s rigid pursuit of architecture, constantly transcending service quality and the design innovation, and its contribution to modern urban construction. It is our sincere expectation that BIAD UFo will serve as the connection between the past and the future from a new starting point, marching towards new brilliance.

As a leading international architectural office, OMA focus on architectural design, urban planning and cultural analysis. OMA has been committed to design and construction and planning for more than 30 years. Led by seven partners, OMA has offices in Rotterdam, New York, Beijing, and Hong Kong, and will set up a new one in Doha. It has around 340 employees.

GEHRY + UFo

盖里事务所是一家提供全方位服务的建筑事务所,在学术建筑、博物馆、电影院、表演和商业建筑以及总体规划项目的设计和建设上拥有广泛的国际经验。盖里事务所成立于1962年,坐落在加州洛杉矶,目前拥有约120名员工。事务所承接的每个项目均由弗兰克·盖里亲自设计,辅以事务所广泛资源以及事务所资深合伙人和员工的经验支持。作品常见于世界上多种报纸和期刊,并在全球主要博物馆中展览。

Tensho Takemori, 加利福尼亚州注册建筑师,盖里事务所合伙人。
盖里事务所与UFo一起合作了北京的中国国家美术馆投标竞赛。此次竞赛包括三个阶段,双方都做付出了大量的工作。UFo在帮助我们理解中国和美术馆新项目的意图方面起了很大作用。对我们而言,UFo是我们发展设计的好顾问,帮助我们的设计理念体现中国的特色和灵魂。在最后的竞赛阶段,我们用短短两周时间,合作建造了一个一比一尺度的幕墙样板段。我相信整个投标过程对双方团队都很有实践意义,我们也期待日后再次合作。

中国国家美术馆

5.2

National Art Gallery of China

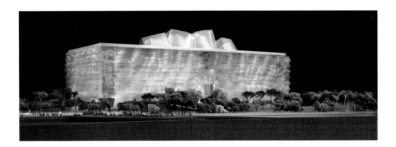

Tensho Takemori, registered and licensed in the State of California. He is a partner of Gehry Partners.
Gehry Partners worked in conjunction with BIAD UFo studio on our competition submission for the National Art Museum of China in Beijing. The competition consisted of three separate rounds, and a significant amount of office work by both sides. BIAD UFo was instrumental in helping us understand China, and the goals the museum should have for their new project. They acted as a sounding board for us as we developed our design, instilling the design with a Chinese soul. In the last competition phase, we collaborated in constructing a full-scale mock-up of the proposed façade design, which was completed in just two weeks. I believe the overall experience was educational for both sides, and we look forward to possible collaborations in the future.

Gehry Partners LLP is a full service architectural firm with extensive international experience in the design and construction of academic institutions, museums, cinemas, performance spaces, commercial buildings, and master planning projects. Founded in 1962 and located in Los Angeles, California, Gehry Partners currently has a staff of approximately 120 people. Every project undertaken by Gehry Partners is designed personally and directly by Frank Gehry, and supported by the extensive resources of the firm and the extensive experience of the firm's senior partners and staff. The work of Gehry Partners has been carried in national and international newspapers and magazines, and has been exhibited in major museums throughout the world.

ZHA + UFo

扎哈·哈迪德建筑师事务所的创始人哈迪德女士曾于2004年荣获被誉为建筑界诺贝尔奖的普利兹克建筑奖。扎哈·哈迪德筑师事务所一直引领着全球建筑界的开拓性研究与设计探讨。通过与众多领先企业的合作,事务所进一步丰富了业务的多样性和自身知识,同时运用先进的技术以实现其流畅、动感与多元化结构的建筑理念。

大桥谕,扎哈·哈迪德建筑事务所合伙人,北京工作室负责人。

祝贺 BIAD UFo工作室成立十周年!我代表扎哈·哈迪德建筑师事务所,非常感谢UFo把我们当作"同行者"。我想特别感谢邵伟平先生、李淦先生以及他们的优秀团队,还有BIAD的所有人。我们有机会亲密合作并完成了标志性的银河SOHO项目,这个项目代表了北京,也是对当代建筑的回应。我们祝愿UFo今后一路迈向成功,也希望我们能有机会再度合作。

银河SOHO

5.3

Galaxy SOHO

Satoshi Ohashi, associate of Zaha Hadid Architects, director of Beijing office.
Congratulations to BIAD UFo on their 10th Anniversary! On behalf of Zaha Hadid Architects, we thank you very much for inviting us to this special portfolio of UFo-Peers. I would like to especially thank Mr. Shao Weiping and his gifted studio, Mr. Li Gan and his team and all of BIAD. We had an opportunity to work very closely together to achieve the iconic Galaxy SOHO Project, representing Beijing and responding to contemporary architecture. We wish the best of success to all of you in the future and our next adventure.

Zaha Hadid, founder of Zaha Hadid Architects, was awarded the Pritzker Architecture Prize (considered to be the Nobel Prize of architecture) in 2004. Zaha Hadid Architects has been a global leader in pioneering research and design investigation. Collaborations with corporations that lead their industries have advanced its business diversity and knowledge, whilst the implementation of state-of-the-art technologies has promoted the realization of fluid, dynamic and therefore diversified architectural structures.

KPF + UFo

KPF在全球设有六处办公室,拥有600多名员工,旨在为全球顶尖客户提供全套建筑设计服务。在30多年的历史中,九次获得美国建筑师协会年终最佳设计奖。建筑类型包括办公、酒店、住宅、学校、医院、交通和综合体项目等,遍布全球至少35个国家。

穆英凯,KPF事务所总监、合伙人。

邵韦平是我的大学同班同学。我俩经常在一起,不仅因为我们是当年班里最高的两个人,还因为我们在生活和建筑设计上拥有许多共同的价值观。我依然记得我们在设计工作室里合作一个位于滨海城市的高层双塔银行项目时的场景。在同济教学楼的地下室,我负责拍摄模型,他负责将胶卷洗成黑白照片。尽管这些照片随着时间的流逝已变得模糊,但30年前的记忆却清晰地留在我的脑海中。在国外学习和工作多年之后,几年前我有幸与同事再次拜访邵韦平和他的工作室。他向我展示了北京机场三号航站楼、凤凰卫视总部等项目,分享了与多个国际顶尖建筑事务所合作的经验。当我在他的工作室第一次看到凤凰卫视总部模型——莫比乌斯环的设计造型时,我便惊奇于它的流线造型是如何适应其复杂的电视总部功能的。而当我看到完成的建筑时,不由对复杂的建筑细节和优美的结构空间表示由衷的赞叹。正如邵韦平团队的名字——UFo所暗示的,他们已进入了自由思考的境界,在北京创造了建筑的奇迹。位于北京CBD的Z15项目是KPF与邵韦平领导的团队正在合作的项目,我们一起设计了北京第一高楼。通过我们的紧密合作,邵和他的团队已证明了他们不仅是优秀的合作者,也是富有创新的建筑师。他的团队已掌握了高难度的塔楼技术,并能根据客户对于塔楼的特殊要求进行优化改善。此次合作使我仿佛回到了30年前我们合作设计的那一刻。虽然我们在空间上相距甚远,但每12小时循环同步工作如同莫比乌斯环,两地办公室无缝同步地为同一项目而努力。

中国尊

5.4

China Zun

Mu Yingkai, director and partner of the KPF.

Shao Weiping was my college classmate. We often accompanied each other, not only because we were the two tallest guys in the class, but also because we shared lots of common values on life and architectural design. I can still remember the scene when we worked together in the design studio for the project, which was a tall twin-tower bank located in a coastal city. In the basement of the teaching building of Tongji University, I photographed models, and he developed the films into black-and-white photos. Although the photos became blurred as time went by, the memories 30 years ago remain clear in my brain. After many years studying and working abroad, I got the chance to visit Shao Weiping and his studio again with my colleagues a few years ago. He showed me the projects of Terminal 3 of Beijing Airport, the headquarters of Phoenix TV, etc., and shared the experience of cooperating with multiple international top architectural firms. When I first saw the model of the headquarters of Phoenix TV, that was the design modeling of the Mobius Strip in his studio, I was surprised at how the streamlined modeling was adapted to the complex functions of the TV headquarters. But when I saw the completed architecture, I expressed my sincere admiration for the complex architectural details and the beautiful structural space. Just as the name of the Shao Weiping team (UFo) suggests, they have entered into the environment with free thinking, and created an architectural miracle in Beijing. The Z15 project located in the Beijing Central Business District is the project that KPF is working on with the team led by Shao Weiping, and we have designed the tallest building in Beijing together. With our close cooperation, Shao and his team have proved that they are not only great cooperators, but also innovative architects. His team has mastered the tower technology and his team will optimize and improve the proposal according to the special requirements of the customers for towers. And the on-going cooperation with Shao reminds me of the joint cooperative design 30 years ago. Far apart, we work as circularly and synchronously as the Mobius Strip every 12 hours, and the offices in the two places design for the same project synchronously and seamlessly.

KPF has six offices worldwide and has more than 600 employees, aiming to provide the full range of architecture design services for premium customers around the world. Through a history of more than 30 years, KPF has won the annual best design award of the American Institute of Architects on nine occasions. Its architectural types include offices, hotels, houses, schools, hospitals, traffic, complex projects and so on, which are scattered in at least 35 countries in the world.

HHD_FUN + UFo

华汇设计（北京）是一个由建筑师、设计师、程序设计师等组成的致力于设计和研究的事务所。参数化设计及可持续发展是其主要研究方向。他们将不同领域的知识创造性地带入建筑设计，希望借由不同的设计方法和设计过程创造出有别于常规的"意料之外"的设计。数学、几何学、算法技术、建筑信息模型、电子学、人工智能学等都是他们涉猎的领域。

王振飞，华汇设计（北京）创始人、主持建筑师，荷兰贝尔拉格学院硕士学位。

第一次到UFo工作室进行交流是在2009年，那时我刚回国建立公司不久，受邀去介绍我们刚做的小项目以及在荷兰留学的成果。那时参数化还是个时髦的新词，在中国刚刚开始流行，在实际项目上鲜有应用。而当时UFo工作室就已经在使用DP进行凤凰中心的BIM设计了，这让我非常惊讶，同时也非常兴奋。作为先行者，UFo工作室对前沿技术的大面积应用，不仅让参数化这个词语在中国找到了语境，同时也让我看到了这套方法在中国应用的广阔前景。那次交流中提到了我们之前的一些几何及算法上的研究，大家都表现出了很大的兴趣，也为之后我们的合作打下了基础。真正的合作是从妫河建筑创意园规划国际竞赛项目开始的，我们作为算法顾问为这个项目提供了一些技术支持。在合作的过程中，UFo工作室的建筑师们对参数化所表现出的热情令我们十分激动，我们在提供自己擅长领域的技术支持的同时，也向各位建筑师学到了很多。正因为有了双方知识的碰撞，才使得最终的成果对双方来说都是出乎意料的，这也是合作的真正意义所在。最近这些年，UFo工作室已经将包括凤凰中心在内的多个项目的成功经验广泛推广，并取得了显著的成绩，这使我们这些年轻建筑师为之鼓舞，真诚希望今后有更多的机会能够进行多方面的合作。

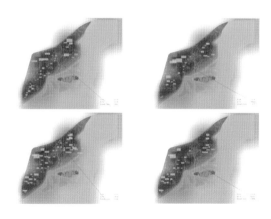

妫河建筑创意园规划国际竞赛方案

5.5

the international competition project for the planning of the Gui River Architectural Innovation Park

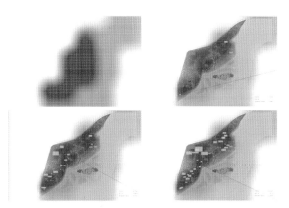

Wang Zhenfei, founder and principal architect of HHD_FUN. He received a Master's Degree from the Berlage Institute in Holland.

The first time I went to UFo studio was for an exchange back in 2009, when I just got back to China and founded my own office. I was invited to share a small project we just did and what I'd studied in Holland. At that time parameterization was still a novel thing, just began to get popular in China, not to mention being applied in actual projects. That's why I felt so surprised and excited when I learnt that UFo studio was already using DP for the BIM design of the Phoenix Center. As a forerunner, UFo has widely applied the state-of-art technology, not only fostering a context for parameterization in China, but also showing me a promising prospect of this method in China. I shared some of our previous researches on geometry and algorithms during that visit, they showed great interest in them, and that became a good start for our future cooperation. The actual cooperation began with the international competition project for the planning of the Gui River Architectural Innovation Park. We provided some technical support for the project as consultant on algorithms. During the cooperation, we were very excited at the enthusiasm the architects of UFo studio showed in the parameterization - we were not only providing technical support in the fields we were good at, but also learning from them. Thanks to the knowledge collision between two sides, the final result was unexpectedly good for both sides. This is also the true meaning of cooperation. In recent years, UFo has widely disseminated the successful experiences of many of their projects including the Phoenix Center, and has seen remarkable results. This greatly encourages young architects like us, and we sincerely hope that there will be more opportunities to work together in the future.

HHD_FUN is an office consisting of architects, designers, programmers, etc. and devotes itself to design and research. Parametric Design and sustainable development are their major research directions. HHD_FUN brings knowledge of different fields creatively into architectural design, and expects to create unexpected designs different from the conventions with different design methods and design processes. Mathematics, geometry, algorithmic technique, building information modeling, electronics, artificial intelligence, etc. are just some of the fields in which they dabble.

SAKO + UFo

SAKO建筑设计工社成立九年来,已建成70余项工程,业务涉及中国、日本、韩国、蒙古和西班牙。设计以建筑设计及室内设计为中心,涉及平面设计、标识、家具、景观及城市规划等广大范围。团队一直坚持致力于发展具有"中国特色"的建筑设计作品。

迫庆一郎,SAKO建筑设计工社主持建筑师。

十年来中国的建筑界发生了迅猛的变化。在与UFo共同经历的这十年当中,我们经受了同样的历练,对UFo有着一种在同一战壕里奋斗的战友般的亲密感。我至今还清晰地记得UFo合伙人刘宇光先生来我公司时的情景。听完他对凤凰中心项目的介绍,我们对该建筑设计质量的高端和其所具有的前瞻性感到惊讶。同时,作为一家外国设计公司,能参与这样一个给首都环境增添崭新一面的公共建筑项目的设计,我们深感荣幸。邵韦平先生领导的UFo团队与我们的合作过程,可以说是一个共同切磋、琢磨的创作过程。同行之间的专业交流更加激发了我们的创意。现在想想,整个过程都是让人兴奋的体验。期待今后我们双方都能继续为中国建筑界的繁荣作出贡献。

凤凰中心室内设计

5.6

Phoenix Center interior design

Keiichiro Sako, principal of SAKO Architects.
During the past decade, Chinese architecture has also changed rapidly. During that period shared with UFo, we have the same experience and have toughened up, and we always have a sense of intimacy with them, like the comrades-in-arms struggling in the same trench. Today, I still clearly remember the scene when the partner of UFo, Mr. Liu Yuguang, came to my corporation. After the introduction of the Phoenix Center project, we were surprised at the high-end design quality of the architecture and the perceptiveness that the architecture had. Meanwhile, as a foreign design corporation, we felt greatly honored to take part in a public architectural project that would add new insights for the capital environment. The process, in which the UFo team led by Mr. Shao Weiping cooperated with us, was a production process of mutual improvement. Meanwhile, the professional communication among peers motivated our originality even more. Now thinking about this, I feel that the whole process was an exciting experience. I expect that both of us can continue to contribute to the prosperity of Chinese architecture in the future.

Beijing Sako Architectural Design Consultant Inc has constructed more than 70 projects since it was established 9 years ago. The business involves China, Japan, Korea, Mongolia and Spain. The designs center on architectural design and interior design, involving a large design portfolio, including graphic design, signs, furniture, landscape, and urban planning, etc. The team has always been committed to developing architectural design works with Chinese characteristics.

KCAP + UFo

KCAP建筑与规划事务所是一家总部位于荷兰的国际化公司，专注于建筑与规划设计，在鹿特丹、苏黎世及上海都均设有办公室。KCAP以能够应对建筑设计与城市规划之间的边缘领域，进行独特的概念和优秀的设计而享有盛名，创造了一系列具有强烈个性和创造性的怡人的城市环境。

Ruurd Gietema, KCAP事务所合伙人，建筑师，城市规划师。

无论是在荷兰或是其他地方，KCAP都依靠自身的设计和研究能力取得了傲人的成绩。然而，当前一个逐步明显的事实是，除去长期的规划和研究能力，通过综合的方法来改变建筑环境还要求不同的、更直接的、实际灵活的能力，一种"放手去做，在过程中学习"的方法。UFo设计团队以他们的先锋精神，一直在做这样的事情。例如，对于如何在重要关头处理棘手问题，以及作为设计师如何寻找到新的方法并使其合理化，这些方面UFo团队都能给我们以非常鼓舞人心的启发。我们希望共同见证未来的十年！

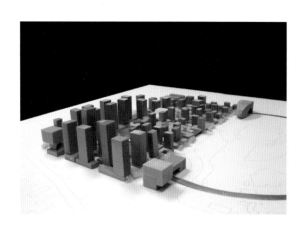

北京奥体文化商务区城市设计

5.7

Beijing Olympic Cultrual Commercial Zone Urban Design

Ruurd Gietema, architect / urban planner and partner in KCAP.
In the Netherlands and Europe KCAP, but also other offices, designs and research led to astonishing results and bravura. However what has become obvious in recent years is that despite all of our long-term planning and research, changing the built environment in a comprehensive way requires a different, more direct and practical dynamic. It requires a more 'just-do-it-and-learning-in-the-process' approach. The designers of BIAD UFo, with their pioneering spirit, have been doing this all along. As an example of how to engage with larger issues at stake and in how to keep finding new ways to legitimize themselves as designers, BIAD UFo are a great inspiration. We hope to witness even more from them the coming ten years.

KCAP is an internationally operating Dutch design firm specializing in architecture and urbanism, with offices in Rotterdam, Zurich and Shanghai. KCAP is reputed for realizing unique concepts and fascinating designs on the fringe between architecture and urbanism. KCAP buildings and master plans demonstrate a track record of creating appealing urban environments with strong individuality.

LAUR Studio + UFo

LAUR Studio是一家从事城市设计与景观设计的专业工作室，合伙人先后在国内外接受严谨的建筑学、景观设计学教育，拥有对不同尺度、不同类型的城市设计和景观设计的丰富经验。在国际前沿的"景观都市主义"的研究支持下，工作室试图在实践与理论之间建立一种互动性的创新对话，探寻在理念上创新、在经济上可行，同时又具有环境责任感的解决方案。

刘东云，博士，LAUR Studio合伙人、主持设计师。
我们先后参与了UFo的多项工程设计项目。在与UFo共同奋斗的日日夜夜里，给我感触最深的是UFo所具有的国际化视野和对细节完美的不懈追求。在邵总的主持下，UFo正成为国际一流的建筑设计团队。与在当今中国经济快速发展背景下进行创作的其他建筑师相比，UFo的设计作品为中国建筑界注入了一股新鲜的血液。UFo对中国建筑界的贡献不仅在于对形式、空间和结构的有趣诠释，还在于他们所提出的当代设计思维与实践模式。这种模式体现为少数设计原则，而这些设计原则是建立在独具匠心的设计研究之上的，并涵盖了科学分析、数学论证、仿生演绎等。这些设计原则在解决工程问题的同时，也塑造了形态特征，这就是自由曲线的流动、组织构成的形式及结构自身的逻辑运动。

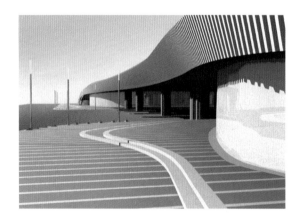

重庆江北国际机场东航站区景观设计

5.8

Chongqing Jiangbei International Airport, East Terminal and Facilities

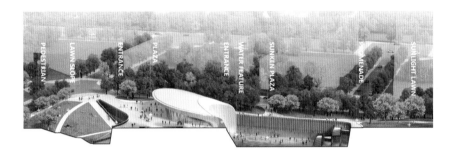

Liu Dongyun, doctor, partner & principal architect of LAUR Studio

We have participated in several UFo design projects. During the days working together with UFo, what struck me most is their international vision and the persistent pursuit for perfection of details. UFo is becoming an internationally renowned architectural design team under the leadership of Mr. Shao. Compared with other architects who are practicing under the background of rapid economic development in China nowadays, the design works of UFo are proven to be a fresh blood into Chinese architecture. They contribute not only an interesting explanations on the form, space and structure, but also an insight into the contemporary design thinking and practice patterns it has developed. This pattern is represented by a few design principles, which are established on ingenious design research, and cover scientific analysis, mathematical arguments, bio-mimetic deduction, etc. The design principles can shape morphological characteristics while solving engineering problems. The principles are the fluxion of free-form curves, the form of organization structure, and the logical movement of the structure.

LAUR Studio is a professional urban design and landscape design studio. The partners have received rigorous education in architecture and landscape architecture successively at home and abroad, having rich experiences in different scales and types of urban design and landscape design. Under the support of study on the leading "Landscape Urbanism", the studio attempts to establish an interactional and innovative dialogue between practice and theory, and to seek a solution that is ideologically innovative, economically feasible and environmentally responsible.

BIAD 结构设计 + UFo

BIAD第四设计院结构团队现有43人,在机场航站楼、超高层建筑、城市综合体、复杂形体建筑等工程的结构设计方面具有丰富的创造性实践经验。

束伟农,BIAD公司副总工程师,兼四院结构设计总监,教授级高工,一级注册结构工程师。

转眼间,四院结构与UFo一起相伴走过了十年。通过北京出版物流发行中心、凤凰中心、银河SOHO、奥南、印度使馆等项目的合作,我们相互信任,并建立了友谊。UFo朝气蓬勃,怀着建筑师的梦想,一路前行。在合作中我们能感受到UFo勇于追求的决心和不懈努力的信心。人因梦想而伟大,更因行动而成功,不管梦想能否实现,追求的过程都是快乐的。凤凰传媒中心的合作,让我们强烈地感受到这一点。建筑与结构专业的沟通相当重要,UFo工作室拥有良好的沟通氛围。UFo工作室的许多项目,四院结构在建筑概念阶段就已经参与其中,因此在设计过程中能够把握建筑师的设计意图,使建筑与结构能够相互协调。四院结构与UFo工作室合作有苦也有甜,虽然我们也会因为工作而进行争论,但目的只有一个,就是把工程完成好。因此,专业间相互理解就显得非常重要,磨合的过程就是相互理解的过程。最后,衷心祝愿UFo在未来的探索之路上再创辉煌。

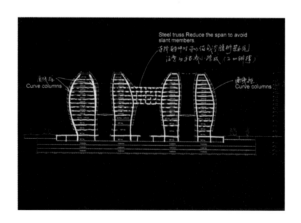

5.9

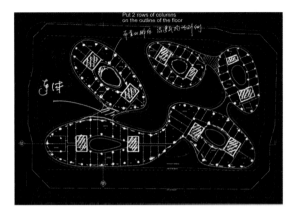

Shu Weinong, deputy chief engineer of BIAD, professorate senior enginee, state first class registered structural engineer.

It has been 10 years since the BIAD Design Department No. 4 Structural Team collaborates with UFo. We have built mutual trust and friendship during the cooperation of projects such as Beijing Publication Logistics Center, Phoenix Center, Galaxy SOHO, Olympic South (Culture Zone, Business Park and Public Space), Chinese Embassy in India, etc. On the 10th anniversary of the establishment of UFo, I would like to share some thoughts on our collaboration on behalf of the structural team. UFo is a vigorous team. It passes along the way with dreams of architects. All of us can feel the determination of its bravery, confidence and passion in the course of cooperation. people are great because of their dreams, but are successful because of their actions. No matter whether the dreams come true, the pursuit is always joyful. Our cooperation on the Phoenix Media Center makes us strongly feel that way. UFo studio also has a sound communication atmosphere. Professional communication between the architectural and the structural disciplines is very important, and for many of UFo projects, our structural team has already got involved at the conceptual phase. In this way, during the design process we could control the design intentions of the architects and coordinate the architecture and the structure. UFo respects mutual understandings between different disciplines. The cooperation between the structural team of No.4 Division and UFo is bittersweet; we may have arguments, but only to one goal: to complete the projects well. Therefore, it's very important for different disciplines to understand each other, and the running-in process is exactly the process of understanding.

There are 43 people in the structural team of BIAD No.4 design division. The team has many experiences in the structural design of projects such as airport terminals, super high-rise buildings, urban complexes, and complicated shape architecture, etc.

Speirs & Major + UFo

Speirs & Major事务所是一家知名设计公司,致力于通过使用灯光来提升人们对视觉环境的体验。业务范围涵盖不同的类型和规模,包含建筑、战略宣传以及产品创新设计。事务所因其帮助并提高了全球的照明设计专业意识而享有盛誉。事务所现有员工30名,他们来自建筑、艺术、室内设计、照明、平面设计、戏剧等领域。在英国的伦敦以及爱丁堡设有办公室。

基思·布拉德肖,Speirs & Major事务所主管,英国皇家艺术学会成员。
Speirs & Major事务所已与BIAD UFo合作了多个项目,这些项目是成功且富有创造性的。我们很荣幸能被选中参与合作北京首都国际机场T3航站楼、凤凰国际传媒中心、CBD中央公园以及奥体南侧中央公园等项目。通过与BIAD UFo的合作,我们共同创造了一些具有个性的项目,他们使项目有了独特性、创新性以及新颖的品质,使项目有了标志性。

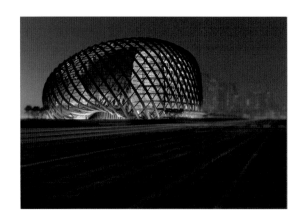

凤凰中心照明设计概念

5.10

Phoenix Center Lighting Concept

Keith Bradshaw, principal of Speirs + Major, fellow of the RSA.
Speirs + Major have worked together on a variety of projects with BIAD UFo, and found the projects to be successful and creative. We are honored to have been selected to collaborate on projects such as: T3 Terminal of Beijing Capital International Airport, Phoenix International Media Center, CBD Central Park and the Olympic South Central Park. Throughout our time working with BIAD UFo, we have jointly produced projects that we believe have individual characteristics, and they create projects of a distinctive, innovative and original quality, enabling our projects to have a symbolic state.

Speirs + Major is a well-known, award-winning design firm that uses light to enhance the experience of the visual environment. Its work is wide-ranging in terms of type and scale and encompasses architecture, strategic branding and innovative product design. We have been credited with helping to raise awareness of the lighting design profession globally. Today the firm employs 30 people drawn from disciplines including architecture, art, interior design, lighting, graphic design and theatre. Its offices are based in London and Edinburgh, UK.

KEREZ + UFo

克里斯蒂安·克雷兹建筑事务所于1993年在瑞士苏黎世成立，由克里斯蒂安·克雷兹教授创办并主持。其建筑作品以充满张力的结构和建筑内部独特的空间设计形成特色，是一家非常具有实验精神的建筑事务所。目前正在进行的主要项目分布在法国、捷克共和国、巴西与中国。

克里斯蒂安·克雷兹, 克里斯蒂安·克雷兹建筑事务所创办人, 苏黎世联邦理工学院建筑学硕士、现任教授。

如果将来, 比方说50年后, 有人回首想看看2014年的建筑, 那他一定得好好看一下中国, 因为在这一年中国城市和景观的变化比世界其他任何地方都来得更巨大、更彻底。但是, 这些变化将仅作为逻辑上的证明, 仅有着经济上的相关性; 若是它们现在不会改变人类以往关于建筑的对话, 那它们将来也不会改变我们对于建筑的理解。建筑最基本的问题是结构, 如何让建筑立起来。通常会依靠立柱和楼板。但是, 在一个被编织得像篮子一样的建筑物里, 那些基本项也就无关紧要了。一想到建筑物的楼板, 人们会期望那是一块平板, 从建筑物的一边延伸到另一边。可是在凤凰中心这些期望要落空了: 楼板在同建筑物外壳分离后变成了独立的存在。看着这幅图时, "立面就是建筑物侧面的覆盖物"这种最基本的理解就显得过时。这里, 建筑物的正面从建筑物的底层开始, 沿着建筑物各面向上攀爬, 一路往上, 最后成为屋顶。

5.11

Christian Kerez, master's degree in architecture, ETH Zurich. Christian Kerez Zurich AG Founder, and professor at ETH Zurich.

If somebody looks back in the future, say in 50 years, and wonders about architecture in the year 2014, he will definitely need to have a good look at China, because in China cities and landscapes changed in that year more dramatically and more drastically than anywhere else. But these changes will have only serve as logistical evidence, will have only an economic relevance, and they will not change our understanding of what architecture can be, if they are not to change the architectural conversations of previous times. The essential issue for a building is about the structure, about how the building stands up. Usually there are columns and slabs, but these basic items would be abrogated in this building, and braided like a basket. If one thinks about a floor in a building, one expects a horizontal slab reaching from one side of the building to the other. But here in Phoenix Center these expectations are questioned once again, as the floors become independent after separation from the shell of the building. It seems that the bottommost understanding of an elevation as the covering of the side view of a building becomes obsolete while looking at this picture. This façade starts underneath the building, climbs up on all sides of the buildings and finally becomes a roof.

Christian Kerez Architectural Firm was established in 1993, in Zurich, Switzerland by Professor Christian Kerez. The architectural works are rather unique with the tension between structures and the particular characteristics of the interior space. Their current projects are mainly located in France, Czech Republic, Brazil and China.

数据库

Database

项目信息 6.1
参展信息 6.2
摄影师信息 6.3
团队成员 6.4

Projects 6.1
Exhibitions 6.2
Photo Credits 6.3
UFo Members 6.4

6.1 项目信息
(按设计开始时间排序)

北京CBD核心区详细规划
项目地点: 北京CBD
规模: 1 500 000平方米
业主: 北京CBD管委会
设计时间: 2003
设计指导: 邵韦平、傅克诚
方案设计: 刘宇光、刘延川、孙纲、王鹏、蔡明、朱江、张士伟

北京世纪华侨城旅游主题社区规划和一期住宅设计
项目地点: 北京朝阳区小武基路世纪华侨城旅游主题社区
规模: 350 000平方米
业主: 北京世纪华侨城实业有限公司
设计时间: 2003—2004
施工时间: 2004—2005
设计指导: 邵韦平
方案设计: 刘宇光、郝亚兰、陈颖、陈莹
设计总负责人: 刘杰、刘宇光、郝亚兰
建筑师: 金国红、陈颖、范楷、杨坤、肖立春、蔡明、赵静、李树栋、陈莹
结构设计/设备设计/电气设计: BIAD 4所

中石油总部
项目地点: 北京市东城区
规模: 200 838平方米
业主: 中国石油集团
设计时间: 2003—2006
施工时间: 2006—2008
设计指导: 刘力、邵韦平
方案设计: 王蔚 (BIAD 4所)、刘宇光、金国红 (BIAD 2所)、吴晨 (TFP)

华侨城集团北京总部
项目地点: 北京世纪华侨城主题社区欢乐谷公园
规模: 4 282平方米
业主: 北京世纪华侨城实业有限公司
设计时间: 2005
施工时间: 2005—2006
设计指导: 邵韦平
方案设计: 刘宇光、李淦、陈颖、吴晶晶
设计总负责人: 李淦
建筑师: 陈颖

Projects (By design begins chronologically order)

Detailed Plan for Beijing CBD Core Area
Location: CBD, Beijing
GFA: 1,500,000 m²
Client: Beijing CBD Committee
Design period: 2003
Director: Shao Weiping, Fu Kecheng.
Schematic design: Liu Yuguang, Liu Yanchuan, Sun Gang, Wang Peng, Cai Ming, Zhu Jiang, Zhang Shiwei

Beijing OTC
Location: Touring Theme Community, Xiao Wuji Road, Chaoyang District, Beijing
GFA: 350,000 m²
Client: Beijing OCT
Design period: 2003-2004
Construction period: 2004-2005
Director: Shao Weiping
Schematic design: Liu Yuguang, Hao Yalan, Chen Ying, Chen Ying
Project supervisor: Liu Jie, Liu Yuguang, Hao Yalan
Architect: Jin Guohong, Chen Ying, Fan kai, Yang Kun, Xiao Lichun, Cai Ming, Zhao Jing, Li Shudong, Chen Ying
SMEP design: BIAD Architectural Design Division No.4

Headquarters of China National Petroleum Corporation
Location: Dongcheng District, Beijing
GFA: 200,838 m²
Client: China National Petroleum Corporation
Design period: 2003-2006
Construction period: 2006-2008
Director: Liu Li, Shao Weiping
Schematic design: Wang Wei (BIAD Architectural Design Division No.4), Liu Yuguang, Jin Guohong (BIAD Architectural Design Division No.2), Wu Chen (TFP)

HQ of OTC Group Beijing
Location: Happy Valley In Beijing Century OCT Theme Community
GFA: 4,282 m²
Client: Beijing OCT
Design period: 2005
Construction period: 2005-2006
Director: Shao Weiping
Schematic design: Liu Yuguang, Li Gan, Chen Ying, Wu Jingjing
Project supervisor: Li Gan
Architect: Chen Ying

2003

2005

北京图书大厦二期
项目地点: 北京市西城区西长安街十七号
规模: 43 122平方米
业主: 北京图书大厦有限责任公司
设计时间: 2005—2006
施工时间: 2007—
方案设计: 邵韦平、李淦、刘宇光
设计总负责人: 邵韦平、李淦、王明霞
建筑师: 蔡明、陈颖、范楷、晓帆、丁明达、温琳琳、郝一涵
结构设计/设备设计/电气设计: BIAD 4所

北京出版发行物流中心规划
项目地点: 北京市通州区台湖镇
规模: 289 596平方米
业主: 北京发行集团
设计时间: 2005—2006
施工时间: 2006—2007
设计指导: 邵韦平
方案设计: 李淦、刘延川、刘宇光
设计总负责人: 李淦
建筑师: 陈颖、王天易
结构设计/设备设计/电气设计: BIAD 4所
景观设计: 房木生景观设计（北京）有限公司

北京国际图书城
项目地点: 北京市通州区台湖镇
规模: 76 300平方米
业主: 北京发行集团
设计时间: 2005—2006
施工时间: 2006—2007
设计指导: 邵韦平
方案设计: 李淦、刘宇光、陈颖
设计总负责人: 李淦
建筑师: 陈颖、苏波、晓帆、李齐颖
结构设计/设备设计/电气设计: BIAD 4所
展厅室内设计: BMA

北京出版发行物流中心酒店
项目地点: 北京通州区台湖镇
规模: 36 736平方米
业主: 北京发行集团
设计时间: 2005—2006
施工时间: 2006—2007
设计指导: 邵韦平
方案设计: 刘延川、李淦、刘宇光、国夫
设计总负责人: 刘延川、李淦
建筑师: 国夫、顾知春、晓帆、吴晶晶、李齐颖
结构设计/设备设计/电气设计: BIAD 4所
室内设计: BMA

Phase 2 of Beijing Book Building
Location:NO.17, West Changan Road, Xicheng District, Beijing.
GFA: 43,122 m^2
Client: Beijing Book Building Co., Ltd.
Design period: 2005-2006
Construction period: 2007- present
Schematic design: Shao Weiping, Li Gan, Liu Yuguang
Project supervisor: Shao Weiping, Li Gan, Wang Mingxia
Architect: Cai Ming, Chen Ying, Fan Kai, Xiaofan, Ding Mingda, Wen Linlin, Hao Yihan
SMEP design: BIAD Architectural Design Division No.4

Planning of Beijing Publication Logistics Center
Location: Taihu Town, Tongzhou District, Beijing
GFA: 289,596 m^2
Client: Beijing Publishing Group
Design period: 2005-2006
Construction period: 2006-2007
Director: Shao Weiping,
Schematic design: Li Gan, Liu Yanchuan, Liu Yuguang
Design principals: Li Gan
Architect: Chen Ying, Wang Tianyi
SMEP design: BIAD Architectural Design Division No.4
Landscape design: Farmerson Architects

Beijing International Book Mall
Location: Taihu Town Tongzhou District, Beijing
GFA: 76,300 m^2
Client: Beijing Publishing Group
Design period: 2005-2006
Construction period: 2006-2007
Director: Shao Weiping
Schematic design: Li Gan, Liu Yuguang, Chen Ying
Project supervisor: Li Gan
Architect: Chen Ying, Su Bo, Xiao Fan, Li Qiying
SMEP design: BIAD Architectural Design Division No.4
Interior design: BMA

Beijing Publication Logistics Center Hotel
Location: Taihu Town, Tongzhou District, Beijing
GFA: 36,736 m^2
Client: Beijing Publishing Group
Design period: 2005-2006
Construction period: 2006-2007
Director: Shao Weiping
Schematic design: Liu Yanchuan, Li Gan, Liu Yuguang, Guo Fu,
Design principals: Liu Yanchuan, Li Gan
Architect: Guo Fu, Gu Zhichun, Xiao Fan, Wu Jingjing, Li Qiying,
SMEP design: BIAD Architectural Design Division No.4
Mall interior Design: BMA

北京出版发行物流中心物流仓储配送中心
项目地点: 北京市通州区台湖镇
规模: 126 200平方米
业主: 北京发行集团
设计时间: 2005—2006
施工时间: 2006—2007
设计指导: 邵韦平
方案设计: 李淦, 刘宇光, 吴岳
设计总负责人: 吴岳, 李淦
建筑师: 王天易, 贾少昆
结构设计/设备设计/电气设计:BIAD研究所

北京世纪华侨城旅游主题社区幼儿园
项目地点: 北京朝阳区小武基路世纪华侨城旅游主题社区
规模: 2 810平方米
业主: 世纪华侨城实业有限公司
设计时间: 2005—2006
施工时间: 2007—2008
设计指导: 邵韦平
方案设计: 刘延川、刘宇光、李淦、蔡明、吴晶晶
设计总负责人: 刘延川、刘宇光
建筑师: 蔡明、吴晶晶

北京世纪华侨城旅游主题社区二期高层住宅A2-5/6/7
项目地点: 北京朝阳区小武基路世纪华侨城旅游主题社区
规模: 82 746平方米
业主: 北京世纪华侨城实业有限公司
设计时间: 2005—2007
施工时间: 2007—2008
设计指导: 邵韦平
方案设计: 刘宇光、苏波
设计总负责人: 樊则森、王炜
建筑师: 樊则森、王阳
结构设计/设备设计/电气设计: BIAD 10所

中华人民共和国驻澳大利亚大使馆
项目地点: 堪培拉亚拉鲁姆拉区128分区第5地段
规模: 6 858平方米
业主: 外交部行政司
设计时间: 2006
施工时间: 2011—2013
设计指导: 邵韦平
方案设计: 刘宇光、李淦、苏波、刘延川
设计总负责人: 刘宇光
建筑师: 王宇、潘辉、冯冰凌、顾知春、丁明达、国夫、陈颖、蔡明
结构设计/设备设计/电气设计: BIAD 6所

Storage and Distribution Center
Loction: Taihu Town, Tongzhou District, Beijing
GFA: 126,200 m²
Client: Beijing Publishing Group
Design period: 2005-2006
Construction period: 2006-2007
Director: Shao WeipingSchematic design: Li Gan, Liu Yuguang, Wu Yue
Design principals: Wu Yue, Li Gan
Architect: Wang Tianyi, Jia Shaokun
SMEP design: BIAD Research Department

Kindergarten of Touring Theme Community in Beijing OCT
Location: Touring Theme Community, Xiao Wuji Road, Chaoyang District, Beijing
GFA: 2,810 m²
Client: Beijing OCT
Design period: 2005-2006
Construction period: 2007-2008
Director: Shao Weiping
Schematic design: Liu Yanchuan, Liu Yuguang, Li Gan, Cai Ming, Wu Jingjing
Design principals: Liu Yanchuan, Liu Yuguang
Architect: Cai MIng, Wu Jingjing

Housing Design of Touring Theme Community in Beijing OCT (Phase 2)
Location: Touring Theme Community, Xiao Wuji Road, Chaoyang District, Beijing
GFA: 82,746 m²
Client: Beijing OCT
Design period: 2005-2007
Construction period: 2007-2008
Director: Shao Weiping,
Schematic design:Liu Yuguang, Su Bo
Design principals: Fan Zesen, Wang Wei
Architect: Fan Zesen, Wang Yang
SMEP design: BIAD Architectural Design Division No.10

Chinese Embassy in Australia
Location: Yarralumla, Canberra
GFA: 6,858 m²
Client: Ministry of Foreign Affairs of the People's Republic of China
Design period: 2006
Construction period: 2011-2013
Director: Shao Weiping,
Schematic design: Liu Yuguang, Li Gan, Su Bo, Liu Yanchuan
Design principals: Liu Yuguang
Architect: Wang Yu, Pan Hui, Feng Bingling, Gu Zhichun, Ding Mingda, Guo Fu, Chen Ying, Cai Ming
SMEP design: BIAD Architectural Design Division No.6

BIAD休息亭
项目地点：北京市南礼士路62号
规模：35平方米
业主：北京市建筑设计研究院
设计时间：2006
施工时间：2006
方案设计：邵韦平、刘宇光、国夫
建筑师：邵韦平、刘宇光、国夫
结构工程师：朱忠义

北京华侨城社区学校
项目地点：北京朝阳区小武基路世纪华侨城旅游主题社区
规模：12 060平方米
业主：世纪华侨城实业有限公司
设计时间：2006—2007
施工时间：2008
设计指导：邵韦平
方案设计：刘延川、刘宇光、李淦、王天易、吴晶晶
设计总负责人：刘延川、刘宇光
建筑师：吴晶晶、王天易
结构设计/设备设计/电气设计：BIAD 4所
景观设计：美国ATA设计公司中国机构

奥林匹克花园中心区下沉花园中国传统元素、总体规划及1、4、5号院
项目地点：奥林匹克公园中心区
规模：45 000平方米
业主：北京新奥集团有限公司
设计时间：2006—2007
施工时间：2007—2008
方案设计：邵韦平、刘宇光、陈淑慧、国夫、苏波、李淦、刘延川、王宇、陈颖、范楷、晓帆、顾知春
设计指导：朱小地、刘力
设计总负责人：邵韦平、陈淑慧、刘宇光
建筑师：王宇、国夫、苏波、陈颖、范楷、晓帆、王天易
结构设计：BIAD 1所
设备设计：BIAD 9所
电气设计：BIAD 1所
植栽配置：BIAD景观工作室
灯光设计：BIAD灯光工作室

上海世博会主题馆
项目地点：上海浦东世博会园区内
规模：80 000平方米
业主：上海世博会事务协调局
设计时间：2007
设计指导：朱小地、胡越、柴裴义
方案设计：邵韦平、刘宇光、刘延川、李淦、肖立春、苏波、窦志、周力大、朱勇、刘明骏、谢鑫、于波
顾问：姜珺／城市中国
景观设计：美国ATA设计公司中国机构
生态设计：马晓军、焦舰

BIAD Pavilion
Location: No. 62 Nanlishi Road, Beijing
GFA: 35 m²
Client: Beijing Institute Of Architectural Design
Design period: 2006
Construction period: 2006
Schematic design: Shao Weiping, Liu Yuguang, Guo Fu
Architect: Shao Weiping, Liu Yuguang, Guo Fu
Structural engineer: Zhu Zhongyi

School of Beijing OCT
Location: Touring Theme Community, Xiao Wuji Road, Chaoyang District, Beijing
GFA: 12,060 m²
Client: Beijing OCT
Design period: 2006-2007
Construction period: 2008
Director: Shao Weiping
Schematic design: Liu Yanchuan, Liu Yuguang, Li Gan, Wang Tianyi, Wu Jingjing
Design principals: Liu Yanchuan, Liu Yuguang
Architect: Wu Jingjing, Wang Tianyi
SMEP design: BIAD Architectural Design Division No.4
Landscape design: ATA Architects & Planners, Beijing office

Sunken Garden in the Center Area of Olympic Green, Master Plan and Courtyards No.1, 4 and 5
Location: Center Area of Olympic Green
GFA: 45,000 m²
Client: Beijing Inno-Olympic Group Co.ltd.
Design period: 2006-2007
Construction period: 2007-2008
Schematic design: Shao Weiping, Liu Yuguang, Chen Shuhui, Guo Fu, Su Po, Li Gan, Liu Yanchuan, Wang Yu, Fan Kai, Xiao Fan, Gu Zhichun
Director: Zhu Xiaodi, Liu Li
Design principals: Shao Weiping, Chen Shuhui, Liu Yuguang
Architect: Wang Yu, Guo Fu, Su Bo, Chen Ying, Fan Kai, Xiao Fan, Wang Tianyi
Structural design: BIAD Architectural Design Division No.1
Mechanical design: The 9th. Design Dept.
Electrical design: BIAD Architectural Design Division No.1
Greenery design: Landscaping Building Planning Studio, BIAD
Lighting design: Lighting Studio, BIAD

Theme Pavillion, Shanghai World Expo 2010
Location: Shanghai Expo Garden
GFA: 80,000 m²
Client: Coordination of Shanghai World Expo
Design period: 2007
Director: Zhu Xiaodi, Hu Yue, Chai Peiyi
Schematic design: Shao Weiping, Liu Yuguang, Liu Yanchuan, Li Gan, Xiao Lichun, Su Bo, Dou Zhi, Zhou Lida, Zhu Yong, Liu Mingjun, Xie Xin, Yu Bo
Consultant: Jiang Jun/ Urban China
Landscape design: ATA Architects & Planners, Beijing office
Ecological design: Ma Xiaojun, Jiao Jian

朝阳区规划展览馆
项目地点：北京市朝阳公园东北角
规模：13 000平方米
业主：北京市朝阳区人民政府
设计时间：2007
设计指导：邵韦平
方案设计：刘宇光、李淦、王宇、闵盛勇、陈颖、刘延川、顾知春、吴晶晶、杨坤

西单文化广场
项目地点：北京市西城区西单北大街西单文化广场
规模：26 274平方米
业主：北京市西城区人民政府
设计时间：2007—2008
施工时间：2008—2009
方案设计：邵韦平、刘宇光、丁明达、蔡明
设计总负责人：邵韦平、刘宇光，
建筑师：蔡明、陈颖、范楷、晓帆
结构设计/设备设计/电气设计：BIAD 4所
景观设计：美国ATA设计公司中国机构
照明设计：北京见阁照明设计公司

凤凰中心
项目地点：北京朝阳公园
规模：72 478平方米
业主：凤凰卫视有限公司
设计时间：2007—2013
建造时间：2008—2014
方案设计：邵韦平、刘宇光、李淦、肖立春、陈颖、刘延川
设计总负责人：邵韦平、刘宇光、陈颖
建筑师：肖立春、王宇、苏波、潘辉、吴锡、周泽渥、池胜峰、吕娟、王丹妮、郝一涵、顾知春、秦超、孙晓宁、袁大伟
结构设计：BIAD 4所+BIAD复杂结构工作室
设备设计/电气设计：BIAD 4所
方案竞赛合作设计：BIAD 4所
方案竞赛顾问：姜珺/城市中国
参数化顾问：北京数字营国信息技术有限公司

景观设计：景观都市主义工作室
灯光概念设计：Speirs + Major事务所
公共区室内概念顾问：SAKO建筑公社
绿色设计：BIAD绿色建筑研究所
声学设计：BIAD研究所

Chaoyang District Urban Planning Exhibition Hall
Location: Chaoyang District, Beijing
GFA: 13,000 m^2
Client: The People's Government Of Chaoyang District In Beijing
Design period: 2007
Director: Shao Weiping
Schematic design: Liu Yuguang, Li Gan, Wang Yu, Min Shengyong, Chen Ying, Liu Yanchuan, Gu Zhichun, Wu Jingjing, Yang Kun

Xidan Cultural Square
Location: Xidan North street, Xicheng District of Beijing Xidan Cultural Square
GFA: 26,274 m^2
Client: The People's Government of Xicheng District In Beijing
Design period: 2007-2008
Construction period: 2008-2009
Schematic designSchematic design: Shao Weiping, Liu Yuguang, Ding Mingda, Cai Ming, Su Bo
Design principals: Shao Weiping, Liu Yuguang
Architect: Cai Ming, Chen Ying, Fankai, Xiao Fan
SMEP design: BIAD Architectural Design Division No.4
Landscape design: ATA Architects & Planners
Lighting design: Beijing Jiange Lighting Equipment Co.,Ltd.

Phoenix Center
Location: Chaoyang Park, Beijing
GFA: 72,478 m^2
Client: Phoenix Co., Ltd.
Design period: 2007-2013
Construction period: 2008-2014
Schematic designSchematic design: Shao Weiping, Liu Yuguang, Li Gan, Xiao Lichun, Chen Ying, Liu Yanchuan
Design principals: Shao Weiping, Liu Yuguang, Chen Ying
Architect: Xiao Lichun, Wang Yu, Su Bo, Pan Hui, Wu Xi, Zhou Zewo, Chi Shengfeng, Lv Juan, Wang Danni, Hao Yihan, Gu Zhichun, Qin Chao, Sun Xiaoning, Yuan Dawei
Structural design: BIAD Architectural Design Division No.4
MEP design: BIAD Architectural Design Division No.4
Competition cooperator: BIAD Architectural Design Division No.4
Competition consultant: Jiang Jun/Urban China
Parametric consultant: BIMTechnologie

Landscape design: LAUR Studio
Lighting conception design: Speirs and Major Associates
Interior conception design: Sako Architects
Greenery: Green Building Research Dept.
Acoustics design: BIAD Research Dept.

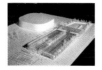  

西双版纳机场航站楼
项目地点: 云南省西双版纳傣族自治州景洪市
规模: 33 200平方米
业主: 云南省机场建设集团
设计时间: 2008
方案设计: 邵韦平、李淦、刘延川、刘宇光、闵盛勇、王宇、吴晶晶
空侧构型及民航工艺设计: 民航新时代机场设计研究院有限公司广州分公司
景观设计: 景观都市主义工作室

奥林匹克花园中心区电话亭
项目地点: 北京奥林匹克公园中心区
规模: 43.2平方米 (共12个)
业主: 北京新奥集团有限公司
设计时间: 2008
施工时间: 2008
设计指导: 邵韦平
方案设计: 刘延川、刘宇光、李淦、苏波、陈颖
设计总负责人: 刘延川
建筑师: 苏波、陈颖
结构设计/设备设计/电气设计: BIAD 1所

金融街G6项目概念设计
项目地点: 北京金融街
规模: 75 000平方米
业主: 中国人民银行总行
设计时间: 2008
方案设计: 邵韦平、刘宇光、李淦、肖立春、苏波、王宇、陈颖、吴晶晶、晓帆、顾知春

北京市建筑设计研究院网页设计
业主: 北京市建筑设计研究院
设计: 李淦、刘延川、晓帆
设计时间: 2008

Xishuangbanna Airport New Terminal
Location: Jinghong, Xishuangbanna Dai Autonomous Prefecture, Yunnan Province
GFA: 33,200 m^2
Client: Yunnan airport group Co.,Ltd.
Design period: 2008
Schematic designSchematic design: Shao Weiping, Li Gan, Liu Yanchuan, Liu Yuguang, Min Shengyong, Wang Yu, Wu Jingjing
Design technology of civil aviation: CAAC New Era Airport Design & Research Institute
Landscape design: LAUR Studio

Telephone Boxes in the Center Area of Olympic Green
Location: Center Area of Olympic Green
GFA: 43.2 m^2 each (12 in total)
Client: Beijing Inno-Olympic Group Co.ltd.
Design period: 2008
Construction period: 2008
Director: Shao Weiping
Schematic design: Liu Yanchuan,Liu Yuguang, Li Gan, Su Bo, Chen Ying
Design principals: Liu Yanchuan
Architect: Su Bo, Chen Ying
SMEP design: BIAD Architectural Design Division No.1

Conceptual Design for G6 on Finance Street
Location: Finance Street, Beijing
GFA: 75,000 m^2
Client: The People'S Bank Of China
Design period: 2008
Schematic designSchematic design: Shao Weiping, Liu Yuguang, Li Gan, Xiao Lichun, Su Bo, Wang Yu, Chen Ying, Wu Jingjing, Xiao Fan, Gu Zhichun

BIAD.com.cn
Client: Beijing Institute of Architectural Design
Design period: 2008
Schematic designSchematic design: Li Gan, Liu Yanchuan, Xiao Fan

银河SOHO
项目地点: 北京市朝阳区东
二环路朝阳门
规模: 330 117平方米
业主: SOHO中国有限公司
设计时间: 2008—2010
施工时间: 2010—2012
方案设计: 英国ZAHA HADID建筑师
事务所
设计指导: 邵韦平
设计总负责人: 李淦、王舒展、
王宇、蔡明
建筑师: 秦超、闫盛勇、吕娟、杨坤、
童佳旎、李玉洁、赵静、卫江、潘辉、
苏波、孙晓宁、王海亮、晓帆、顾知春
结构设计: BIAD 4所
设备设计/电气设计: BIAD 4所+
秦禾国际工作室

珠海市博物馆和城市规划展览馆
项目地点: 珠海市情侣路
与梅华路交汇处
规模: 48 400平方米，其中博物馆
33 000平方米，城市规划展览馆
15 400平方米。
业主: 珠海九洲旅游集团有限公司
设计时间: 2009
方案设计: 邵韦平、李淦、王海亮、
苏波、刘鹏飞、吴晶晶
景观设计: 景观都市主义工作室

建威大厦16层室内设计
项目地点: 北京西城区南礼士路66号
建威大厦16层
规模: 2 000平方米。
业主: 北京市建筑设计研究院
设计时间: 2009
室内设计: 邵韦平、杨坤
机电设计: BIAD 4所+秦禾国际工作室

鄂尔多斯满世广场
项目地点: 鄂尔多斯东胜铁西区
规模: 146 279平方米
业主: 满世房地产开发有限责任公司
设计时间: 2009—2011
施工时间: 2011—
设计指导: 邵韦平
方案设计: 刘延川、李淦、闵胜勇、
吴晶晶、杨坤
设计总负责人: 刘延川、李淦、于辉
建筑师: 欧阳露、温琳琳
结构设计/设备设计/电气设计: BIAD 2所
景观设计: 景观都市主义工作室
灯光设计: BIAD灯光工作室

Galaxy SOHO
Location: Chaoyangmen,
Chaoyang District, Beijing
GFA: 330,117 m²
Client: Soho China Co.,Ltd.
Design period: 2008-2010
Construction period: 2010-2012
Schematic design: Zaha Hadid Architects
Director: Shao Weiping
Design principals: Lee Gan,
Wang Shuzhan, Wang Yu,
Cai Ming
Architect: Qin Chao,
Min Shengyong, Lv Juan,
Yang Kun, Tong Jiani, Li Yujie,
Zhao Jing, Wei Jiang, Pan Hui,
Su Bo, Sun Xiaoning, Wang
Hailiang, Xiao Fan, Gu Zhichun
Structural design: BIAD
Architectural Design Division No.4
MEP design: BIAD Architectural
Design Division No.4

Zhuhai City Museum and the Urban Planning Exhibition Hall
Location: interchange of Qinglv
Road and MeiHua Road in Zhuhai
GFA: 48,400 m² (museum: 33,000
m²; Urban Planning Exhibition Hall:
15,400 m²)
Client: Zhuhai Jiuzhou Tourism Group
Design period: 2009
Schematic design: Shao Weiping,
Li Gan, Wang Hailiang, Su Bo,
Liu Pengfei, Wu Jingjing
Landscape design: LAUR Studio

Interior Design of Canway Building F16
Location: RM.1602, Canway
Buliding, 66 Nanlishi Road, Beijing
GFA: 2,000 m²
Client: Beijing Institute of Architectural Design
Design period: 2009
Interior design: Shao Weiping, Yang Kun
MEP design: BIAD Architectural
Design Division No.4, TSINHE
Design Dept.

Erdos Manshi Square
Location: Dongsheng, Tiexi District, Erdos
GFA: 146,279 m²
Client: Manshi real estate Co.,Ltd.
Design period: 2009-2011
Construction period: 2011-present
Director: Shao Weiping
Schematic design: Liu Yanchuan,
Li Gan, Min Shengyong,
Wu Jingjing, Yang Kun
Design principals: Liu yanchuan,
Li Gan, Yu Hui
Achitect: Ouyang Lu, Wen Linlin
MEP design: BIAD Architectural
Design Division No.2
Landscape design: LAUR Studio
Lighting design: BIAD Lighting Studio

   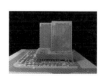

重庆江北国际机场东航站区
及配套设施
项目地点：重庆江北国际机场
规模：T3航站楼351 100平方米，GTC
230 440平方米
业主：重庆机场集团有限公司
设计时间：2009（方案竞赛第一名）
设计联合体：北京市建筑设计研究院
BIAD + 荷兰机场咨询公司NACO
BIAD团队：第四设计所+UFo
UFo方案设计：邵韦平、李淦、闵盛勇、
王海亮、刘延川
景观设计：景观都市主义工作室

北京妫河建筑创意产业园规划
项目地点：北京延庆县妫河北岸
规模：15万平方米
业主：北京建院建筑创意有限公司
设计时间：2009
施工时间：2010—2012
方案设计：邵韦平、刘宇光、李淦、
吴晶晶、刘鹏飞
总图设计：吕娟
设备设计/电气设计：BIAD机电所
景观设计：景观都市主义工作室

北京CBD核心区公共空间
项目地点：北京市朝阳区国贸桥东北角
规模：524 130平方米
业主：北京国际商务中心区
开发建设有限公司
设计时间：2009—2014
施工时间：2011—
方案设计：邵韦平、刘宇光、蔡明、
闵胜勇、李玉洁、石璐、周芸、温琳琳、
孙月恒、马思端、王钊、王宇
设计总负责人：王晓群、刘宇光、蔡明
建筑师：何利文、闵盛勇、吕娟、李玉洁、
石璐、周芸、温琳琳、孙月恒、
马思端、王丹妮、尤明、王钊、王承、卫江、
赵静、王宇、秦超
结构设计：BIAD 9S1工作室
设备设计/电气设计：BIAD 机电所

世博后续开发项目B06超五星级酒店
项目地点：上海市世博园
规模：269 900平方米
业主：上海世博会有限公司
设计时间：2010
方案设计：邵韦平、刘宇光、李淦、
石璐、袁大伟

**Chongqing Jiangbei
International Airport, East
Terminal and Facilities**
Location: Chongqing Jiangbei
International Airport
GFA: Terminal 3: 51,100 m²; GTC:
230,440 m²
Client: Chongqing airport group
Co., Ltd.
Design period: 2009
(The 1st Prize of the Competition)
Schematic design: BIAD + NACO
BIAD team: BIAD Architectural
Design Division No.4 + UFo
UFo tTeam: Shao Weiping, Li Gan,
Min Shengyong, Wang Hailiang,
Liu Yanchuan
Landscape design: LAUR Studio

**Beijing Gui River Architecture
Innovation Park**
Location: Gui River,Yanqing
County, Beijing
GFA: 150,000 m²
Client: Architectural Design
Creation Co.,Ltd.,BIAD
Design period: 2009 (the 1st prize
of the competition)
Construction period: 2010-2012
Schematic design: Shao Weiping,
Liu Yuguang, Li Gan, Wu Jingjing,
Liu Pengfei
Master planning design: Lv Juan
MEP design: BIAD Mechanical and
Electrical Design Dept.
Landscape design: LAUR Studio

CBD Core Area Public Space
Location: Guo Mao Bridge, Chao
Yang District, Beijing China
GFA: 524,130 m²
Client: Beijing international
business district development and
Construction Co., Ltd.
Design period: 2009-2014
Construction period: 2011-present
Schematic design: Shao Weiping,
Liu Yuguang, Cai Ming,
Min Shengyong, Li Yujie,
Shi Lu, Zhou Yun, Wen Linlin,
Sun Yueheng, Ma Siduan,
Wang Zhao, Wang Yu
Design principals: Wang Xiaoqun,
Liu Yuguang ,Cai Ming
Architect: He Liwen, Min
Shengyong, Lv Juan, Li Yujie,
Shi Lu, Zhou Yun, Wen Linlin,
Sun Yueheng, Ma Siduan,
Wang Danni, You Ming,
Wang Zhao, Wang Cheng,
Wei Jiang, Wang Yu, Qin Chao
Structural design:BIAD 9S1 Studio
MEP design: BIAD Mechanical and
Electrical Design Dept.

Post-Expo project: B06 Hotel
Location: Shanghai Expo garden
GFA: 269,900 m²
Client: Expo Shanghai Group
Design period: 2010
Schematic design: Shao Weiping,
Liu Yuguang, Li Gan, Shi Lu,
Yuan Dawei

2010

世博后续开发项目 B14 超高层办公楼
项目地点: 上海市世博园
规模: 184 100 平方米
建筑高度: 239 米
业主: 上海世博会有限公司
设计时间: 2010
方案设计: 邵韦平、李淦、刘宇光、殷霄雷、周泽渥
景观设计: 景观都市主义工作室

奥林匹克公园中心区文化综合区
项目地点: 北京奥林匹克公园中心区
规模: 用地面积 25 公顷
业主: 北京新奥集团有限公司
设计时间: 2010—2011
方案设计: UFo+ KCAP
UFo 团队: 邵韦平、刘宇光、刘鹏飞、尤明

中国驻冰岛大使馆新购馆舍改造项目
项目地点: 冰岛共和国雷克雅未克市
规模: 4 920 平方米
业主: 中华人民共和国外交部
设计时间: 2010—2011
施工时间: 2011—2012
设计指导: 邵韦平
方案设计: 陈颖
设计总负责人: 陈颖
建筑师: 孙晓宁、吴锡
展厅室内设计: 陈颖、吴锡
结构设计: BIAD 4 所
设备设计/电气设计: 秦禾国际工作室

妫河建筑创意产业园接待中心
项目地点: 北京延庆县妫河北岸延庆镇西屯村西南地块
规模: 3 595 平方米
业主: 北京建院建筑设计创意有限公司
设计时间: 2010—2013
施工时间: 2011—2013
设计指导: 朱小地、邵韦平
方案设计: 李淦、刘鹏飞
设计总负责人: 李淦
建筑师: 刘鹏飞、吕娟
殷霄雷、郝一涵
结构设计: BIAD 1S1 工作室
设备设计/电气设计: BIAD 3M1 工作室
室内设计: UFo+北京兆舍建筑设计咨询有限公司+南山工程设计咨询(北京)有限公司
灯光设计: BIAD 灯光工作室
景观设计: 景观都市主义工作室

Post-Expo project: B14 office tower
Location: Shanghai Expo garden
GFA: 184,100m^2
Height: 239 m
Client: Expo Shanghai Group
Design period: 2010
Schematic design: Shao Weiping, Li Gan, Liu Yuguang, Yin Xiaolei, Zhou Zewo
Landscape design: LAUR Studio

Olympic Green Culture Zone
Location: Beijing Olympics park central area
GFA: 25 ha
Client: Beijing Inno-Olympic Group Co.ltd.
Design period: 2010-2011
Schematic design: UFo + KCAP
UFo team: Shao Weiping, Liu Yuguang, Liu Pengfei, You Ming

Reconstruction of the Chinese Embassy in Iceland
Location: Reykjavik, the republic of Iceland
GFA: 4,920 m^2
Client: Ministry of Foreign Affairs of the People's Republic of China
Design period: 2010-2011
Construction period: 2011-2012
Director: Shao Weiping
Schematic design: Chen Ying
Design principals: Chen Ying
Architect: Sun Xiaoning, Wu Xi
Structural design: BIAD Architectural Design Division No.4
MEP design: TSINHE Design Dept.

Show Room of Beijing Gui River Architecture Innovation Park
Location: Southwest of Xitun village, Yanqing town, northern bank of Gui River in Yanqing county, Beijing
GFA: 3,595 m^2
Client: Architectural Design Creation Co., Ltd., BIAD
Design period: 2010-2013
Construction period: 2011-2013
Director: Zhu Xiaodi, Shao Weiping
Schematic design: Li Gan, Liu Pengfei, Lv Juan, Yin Xiaolei, Hao Yihan
Design principals: Li Gan
Architect: Liu Pengfei, Lv Juan, Yin Xiaolei, Hao Yihan
Structural design: BIAD 1S1 Studio
MEP design: BIAD 3M1 Studio
Interior design: UFo + Beijing My House Architecture Design Co.,Ltd. + Nanshan Engineering Consulting (Beijing) Co.,Ltd.
Lighting design: BIAD Lighting Studio
Landscape design: LAUR Studio

联合国总部改建	奥体文化商务园公共空间	中华人民共和国驻印度大使馆	中国尊
项目地点：美国 纽约	项目地点：北京市奥体中心区南侧	项目地点：印度 新德里	项目地点：北京CBD核心区Z15地块
规模：103平方米	规模：30万平方米	规模：13 500平方米	规模：437 000平方米
业主：中国外交部	业主：北京新奥集团有限公司	业主：中国外交部	业主：北京中信和业投资有限公司
设计时间：2011	设计时间：2011—2014	设计时间：2011—2014	设计时间：2011—2014
方案设计：邵韦平、陈卓群	施工时间：2012—	施工时间：2013—	施工时间：2013—

联合国总部改建
项目地点：美国 纽约
规模：103平方米
业主：中国外交部
设计时间：2011
方案设计：邵韦平、陈卓群

奥体文化商务园公共空间
项目地点：北京市奥体中心区南侧
规模：30万平方米
业主：北京新奥集团有限公司
设计时间：2011—2014
施工时间：2012—
方案设计：邵韦平、刘宇光、
郝亚兰、吴晶晶、刘鹏飞
设计总负责人：刘宇光、郝亚兰
建筑师：吴晶晶、吕娟、刘鹏飞、赵静、
王风涛、殷霄雷、陈立维、王钊
结构设计：BIAD 4所
设备设计/电气设计：
BIAD 3M1工作室
室内设计：UFo+北京兆舍建筑设计
咨询有限公司
景观设计：景观都市主义工作室

中华人民共和国驻印度大使馆
项目地点：印度 新德里
规模：13 500平方米
业主：中国外交部
设计时间：2011—2014
施工时间：2013—
方案设计：邵韦平、陈卓群
设计总负责人：邵韦平
建筑师：杨坤、陈卓群、缪一新、
殷霄雷
结构设计/设备设计/电气设计：BIAD 4所
室内设计：UFo+BIAD室内设计一室

中国尊
项目地点：北京CBD核心区Z15地块
规模：437 000平方米
业主：北京中信和业投资有限公司
设计时间：2011—2014
施工时间：2013—
设计指导：朱小地、徐全胜
实施方案设计：
KPF+北京市建筑设计研究院
设计总负责人：邵韦平
设计副总负责人：吴晨
建筑师：奚悦、段昌莉、韩慧卿、陈颖、
苏晨、王骅、王亮、吴懿、朱学晨、朱江、
吕娟、王鸣鸣、刘利、王伟、孙文昊、兰
岚、郝一涵、周泽墨、何淼淼、陈阳、李
晖、陈文刚、佟磊
结构设计：BIAD城市规划与建筑设计所
(中汇国际)+BIAD 4所+BIAD复杂结构所
设备设计：BIAD 2所+BIAD 4所+
城市规划与建筑设计所(中汇国际)
+BIAD绿色研究所
电气设计：秦禾国际工作室+城市规划
与建筑设计所(中汇国际)
BIM：BIAD BIM工作室

The United Nations office renovation
Location: New York, The United States
GFA: 103 m²
Client: Ministry of Foreign Affairs of the People's Republic of China
Design period: 2011
Schematic design: Shao Weiping, Chen Zhuoqun

Olympic South: Culture Zone, Business Park and Public Space
Location: The Olympic sports center of Beijing
GFA: 300,000 m²
Client: Beijing Inno-Olympic Group Co.ltd
Design period: 2011-2014
Construction period: 2012 - present
Schematic design: Shao Weiping, Hao Yalan, Wu Liu Yuguang, Hao Yalan, Wu Jingjing, Liu Pengfei
Design principals: Liu Yuguang, Hao Yalan
Architect: Wu Jingjing, Lv Juan, Liu Pengfei, Zhao Jing, Wang Fengtao, Yin Xiaolei, Chen Liwei, Wang Zhao
MEP design: BIAD 3M1 Studio
Interior design: UFo + Beijing My House Architectural DesignCo.Ltd.
Landscape design: LAUR Studio

Chinese Embassy in India
Location: New Delhi, India
GFA: 13,500 m²
Client: Ministry of Foreign Affairs of the People's Republic of China
Design period: 2011 - 2014
Construction period: 2013 - present
Schematic design: Shao Weiping, Chen Zhuoqun
Design principals: Shao Weiping
Architect: Yang Kun, Chen Zhuoqun, Miao Yixin, Yin Xiaolei
SMEP design: BIAD Architectural Design Division No.4
Interior design: UFo + BIAD Interior Design Studio I

China Zun
Location: Beijing CBD Core Area Z15Plot
GFA: 437,000 m²
Client: CITIC Heye Investment Co.,Ltd.
Design period: 2011-2014
Construction period: 2013 - present
Director: Zhu Xiaodi, Xu Quansheng
Schematic design: KPF, BIAD
Design principals: Shao Weiping
Associate design principle: Wu Chen
Architect: Xi Yue, Duan Changli, Han Huiqing, Chen Ying, Su Chen, Wang Hua, Wang Liang, Wu Yi, Zhu Xuechen, Zhu Jiang, Lv Juan, Wang Mingming, Liu Li, Wang Wei, Sun Wenhao, Lan Lan, Hao Yihan, Zhou Zewo, He Miaomiao, Chen Yang, Li Hui, Chen Wengang, Tong Lei
Structural design: BIAD City plan & Architectural Design Division CDG International + BIAD Architectural Design Division No.4 + Complex Structure Research Division
Mechanical design: BIAD Architectural Design Division No.2 + BIAD Architectural Design Division No.4 + BIAD City plan & Architectural Design Division CDG International + BIAD City plan & Architectural Design Division CDG International
Electrical design: TSINHE Design Dept + BIAD City plan & Architectural Design Division CDG International
BIM: BIM Research Dept

第十三届威尼斯建筑双年展中国国家馆参展设计——序列
项目地点: 威尼斯建筑双年展中国馆
规模: 长23米、宽1米、高1米
业主: 威尼斯建筑双年展
设计时间: 2012
施工时间: 2012
设计: 邵韦平、陈颖、刘鹏飞、周泽渥、袁大伟、王丹妮

巴塘人民小学宿舍
项目地点: 四川省甘孜藏族自治州巴塘县
规模: 6 500平方米
业主: 巴塘县政府
设计时间: 2012—2013
施工时间: 2013—
设计指导: 邵韦平
方案设计: 刘宇光、尤明、王钊
设计总负责人: 刘宇光
施工图设计: 基准方中事务所

CBD-Z14
项目地点: 北京CBD核心区Z14地块
规模: 29 000平方米
业主: 北京CBD管委会
设计时间: 2012—2014
施工时间: 2014—
方案设计: 邵韦平、刘宇光、李淦、陈卓群、朱学晨、Nathanael Paul Maschke
设计总负责人: 邵韦平、李淦
建筑师: 陈卓群、朱学晨
结构设计: BIAD 9S1工作室
电气设计: BIAD秦禾设计所

"Sequence", exhibit of the 13th Venice Biennale
Location: Chinese pavilion, Venice
Scale: 23 m (L) * 1 m (W) * 1 m (H)
Client: the 13th Venice Biennale
Design period: 2012
Construction period: 2012
Schematic design: Shao Weiping, Chen Ying, Liu Pengfei, Zhou Zewo, Yuan Dawei, Wang Danni

Dormitory Design of Batang School Campus
Location: Batang County, Ganzi Tibetan Autonomous Prefecture, Sichuan Province
GFA: 6,500 m²
Client: Batang County Government
Design period: 2012-2013
Construction period: 2013 - present
Director: Shao Weiping
Schematic design: Liu Yuguang, Youming, Wang Zhao
Design principals: Liu Yuguang
LDI: Chengdu JZFZ Architectural Design Co.,Ltd.

CBD-Z14
Location: Beijing CBD Core Area Z14 Plot
GFA: 29,000 m²
Client: Beijing CBD Committee
Design period: 2012-2014
Construction period: 2014 - present
Schematic design: Shao Weiping, Liu Yuguang, Li Gan, Chen Zhuoqun, Zhu Xuechen, Nathanael Paul Maschke
Design principals: Shao Weiping, Li Gan
Architect: Chen Zhuoqun, Zhu Xuechen
Structural design: BIAD 9S1 Studio
Electrical design: BIAD TSINHE Design Dept.

6.2 参展信息
（按参展时间排序）

"城市_新山水"当代城市建筑
艺术邀请展
作品名称：短片 APPROACH TO
地点：重庆城市规划展览馆
时间：2007.12.29—2008.01.29

2007香港·深圳城市\建筑双城
双年展
作品名称：北京国际图书城短片+
凤凰中心+北京CBD核心区城市
设计
地点：香港中区警署及域多利监狱的
古迹建筑群
时间：2008.01—2008.03

"向东方"：中国建筑景观展
作品名称：凤凰中心
地点：意大利罗马二十一世纪
国家艺术博物馆
时间：2011.7.28—10.23

第13届威尼斯国际建筑
双年展中国馆
作品名称：序列
地点：意大利威尼斯
时间：2012.08.29—2012.11.25

Exhibitions
(In Chronological Order)

**Urban_New Spectacle:
Contemporary Urbanism,
Architecture and Invitation
Exhibition**
Title: APPROACH TO (a short film)
Venue: Chongqing Urban Planning
Museum
Time: 2007.12.29-2008.01.29

**2007 Hongkong-Shenzhen
Architecture Biennale**
Title: a short film of Beijing
International Book Mall; Phoenix
Center; Beijing CBD Core Area
Design
Venue: the Central Police Station
Compound, Hongkong
Time: 2008.01-2008.03

**Verso Est. Chinese
Architectural Landscape**
Title: Phoenix International Media
Center
Venue: National Museum of XXI
Century Arts, Roma, Italy
Time: 2011.7.28-10.23

**13th International Architecture
Exhibition, Venice Biennale**
Title: Sequence
Venue: Venice, Italy
Time: 2012.08.29-2012.11.25

建筑中国——100个当代项目
作品名称：CBD核心区地下设施和景观设计、北京首都机场T3航站楼、凤凰中心、中国尊
地点：德国曼海姆莱斯
　　——艾格尔博物馆
时间：2012.08.29—2012.11.25

第13届威尼斯双年展中国馆原初展
　　——上海新天地站
作品名称：序列
地点：上海新天地太平湖，黄陂南路
时间：2013.06.26—2013.09.31

第八届HAY文学艺术节，中国宫
　　——中国当代建筑展
作品名称：凤凰中心、CBD核心区地下设施和景观设计、中国尊
地点：帕拉西奥金塔纳尔宫，塞戈维亚，西班牙
时间：2013.09.21—2013.11.26

数字渗透：数字建筑展
作品名称：序列
地点：北京798艺术区
时间：2013.09.29—2013.10.20

Architecture China-The 100 Contemporary Projects
Title: Beijing CBD Core Underground Public Space and the Central Park; Beijing Capital Airport Terminal 3; Phoenix International Media Center; China Zun
Venue: Reiss - Engelhorn Museen (REM), Mannheim, Germany
Time 2012.09.15-2013.01.13

"Originaire", 13th International Architecture Exhibition, Venice Biennale
Title: Sequence
Venue: Xin Tian Di, Shanghai
Time: 2013.06.26-2013.09.31

The 8th HAY Literature and Art Festival, Palace Of China-Architecture China 2013
Title: Phoenix International Media Center; Beijing CBD Core Underground Public Space and the CBD Central Park; Beijing Capital Airport Terminal 3; China Zun
Venue: Palacio Quintanar, Segovia, Spain
Time: 2013.09.21-2013.11.26

Digital Infiltration
Title: Sequence
Venue: Beijing 798 Art District
Time: 2013.09.29-2013.10.20

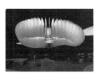

6.3 摄影师信息

特写
2.2 工作模型
傅兴: 36-37

对谈 UFo
谭岚: 52—53

实践
4.11 凤凰中心
傅兴: 90, 92上, 96, 98, 100上, 102—105;
李淦: 95;
周泽渥: 100下
4.12 鄂尔多斯满世广场
傅兴: 106, 108;
李淦: 111
4.13 妫河建筑创意产业园规划
国际竞赛方案
傅兴: 114

4.14 妫河建筑创意产业园接待中心
李淦: 120, 123, 126下, 131;
杨超英: 126上, 129, 130, 132—133
4.15 珠海市博物馆和城市规划展览馆
傅兴: 134
4.17 银河SOHO
王祥东: 144, 146—147上, 148, 150—151
4.18 序列
刘鹏飞: 152, 155
4.21 CBD核心区公共空间
王祥东: 162, 168, 170—171
4.22 奥体文化商务园区公共空间
吴晶晶: 178上;
王祥东: 179
4.23 北京国际图书城
李淦: 181, 185上, 186, 188—189;
傅兴: 183, 187上;
杨超英: 180, 184

4.24 北京世纪华侨城
王祥东: 190, 192—193
4.25 华侨城北京总部
杨超英: 194, 197下;
李淦: 197上
4.26 北京华侨城社区学校
杨超英: 198, 201—203
4.27 中石油总部
傅兴: 204, 207上, 208—209
4.28 朝阳区规划展览馆
傅兴: 210, 213, 214—215上
4.31 奥林匹克公园中心区下沉花园
柳迪: 222, 225, 226上, 227上;
傅兴: 228—229
4.32 北京图书大厦二期
傅兴: 230;
李淦: 232上
4.33 西双版纳机场航站楼
傅兴: 234, 236下
4.34 巴塘人民小学宿舍
傅兴: 238, 244上, 245上

4.35 中国驻澳大利亚大使馆
刘宇光: 246, 250—251;
傅兴: 247
4.36 中国驻印度大使馆
傅兴: 252, 255
4.37 BIAD休息亭
李淦: 258, 260—261

同行者
5.3 ZHA+UFo
王祥东: 272—273
5.11 KEREZ+UFo
KEREZ: 288—289

Photo Credits

Features
2.2 Study model
Fu Xing: 36-37

Interview with UFo
Tan Lan: 52-53

Practice
4.11 Phoenix Center
Fu Xing: 90, 92top, 96, 98, 100top, 102-105;
Li Gan: 95
Zhou Zewo: 100 bottom
4.12 Erdos Manshi Square
Fu Xing: 106, 108;
Li Gan: 111
4.13 Beijing Gui River Architecture Innovation Park
Fu Xing: 114

4.14 Show Room of Beijing Gui River Architecture Innovation Park
Li Gan: 120, 123, 126bottom, 131;
Yang Chaoying: 126top, 129, 130, 132-133
4.15 Zhuhai City Museum and the Urban Planning Exhibition Hall
Fuxing: 134
4.17 Galaxy Soho
Wang Xiangdong: 144, 146-147top, 148, 150-151
4.18 Sequence
Liu Pengfei: 152, 155
4.21 CBD Core Area Public Space
Wang Xiangdong: 162, 168, 170-171
4.22 Olympic South: Culture Zone, Business Park and Public Space
Wu Jingjing: 178top;
Wang Xiangdong: 179
4.23 Beijing International Book Mall
Li Gan: 181, 185top, 186, 188-189;
Fu Xing: 183, 187top;
Yang Chaoying: 180, 184

4.24 Beijing OCT
Wang Xiangdong: 190, 192-193
4.25 HQ of OCT Group Beijing
Yang Chaoying: 194, 197bottom;
Li Gan: 197top
4.26 School of Beijing OCT
Yang Chaoying: 198, 201-203
4.27 Headqarters of China national petroleum corporation
Fu Xing: 204, 207top, 208-209
4.28 Chaoyang District Urban Planning Exhibition Hall
Fu Xing: 210, 213, 214-215top
4.31 Design in the Center Area of Olympic Green
Liu Di: 222, 225, 226top, 227top;
Fu Xing: 228-229
4.32 The Second Phase of Beijing Book Building Project
Fu Xing: 230;
Li Gan: 232top
4.33 Xishuangbanna Airport New Terminal
Fu Xing: 234, 236top
4.34 Dormitory Design of Batang School Campus
Fu Xing: 238, 244top, 245top

4.35 Chinese Embassy In Australia
Liu Yuguang: 246, 250-251;
Fu Xing: 247
4.36 Chinese Embassy In India
Fu Xing: 252, 255
4.37 BIAD Pavilion
Li Gan: 258, 260-261

Cooperators
5.3 ZHA+UFo
Wang Xiangdong: 272-273
5.11 KEREZ+UFo
KEREZ: 288-289

6.4 团队成员

室主任

邵韦平
北京市建筑设计研究院有限公司执行总建筑师，兼UFo建筑工作室主任。教授级高级建筑师，国家一级注册建筑师。中国建筑学会常务理事、中国建筑学会建筑师分会理事长、北京市土木建筑学会理事长、清华大学、中央美术学院、北京建筑大学兼职硕士生导师。1984年毕业于同济大学建筑系，硕士学位。

刘宇光
北京市建筑设计研究院有限公司UFo建筑工作室副主任，兼公司副总建筑师。高级建筑师，国家一级注册建筑师。担任中国建筑学会建筑师分会理事、中国建筑学会建筑师分会数字专业委员会委员、北京市规划学会理事、北京建筑大学兼职硕士生导师。1997年毕业于同济大学建筑系，硕士学位。曾获中国建筑学会青年建筑师奖。

李淦
北京市建筑设计研究院有限公司UFo建筑工作室副主任。高级建筑师，国家一级注册建筑师。1997年毕业于重庆建筑大学建筑城规学院，硕士学位。曾获中国建筑学会青年建筑师奖。

UFo Members

Directors

Shao Weiping
Chief Executive Architect of Beijing Institute of Architectural Design, Director of UFo Studio. Professorate Senior Architect, National First Class Registered Architect. He serves as Executive Director of the Architectural Society of China, President of the Architectural Society of China's Architects Branch, President of the Civil Engineering & Architectural Society of Beijing, and part-time graduate tutor at Tsinghua University, Central University of Fine Arts and Beijing University of Civil Engineering and Architecture. He graduated from the Department of Architecture at Tongji University in 1984 with a Master's Degree.

Liu Yuguang
Deputy Director of UFo Studio, Deputy Chief Architect of BIAD. Senior Architect, National First Class Registered Architect. He serves as Director of the Architectural Society of China's Architects Branch, Director of the Digital Professional Committee of Architectural Society of China Architects Branch, Director of the Planning Society of Beijing, and part-time graduate tutor at Beijing University of Civil Engineering and Architecture. He graduated from the Department of Architecture at Tongji University in 1997 with a Master's Degree. He won the Young Architects Awards of the Architectural Society of China.

Li Gan
Deputy Director of UFo Studio. Senior Architect, National First Class Registered Architect. Graduated from the School of Architecture and Urban Planning at Chongqing Jianzhu University in 1997 with a Master's Degree. He won the Young Architects Awards of the Architectural Society of China.

主任建筑师
陈颖
UFo工作室主任建筑师,高级建筑师,国家一级注册建筑师。2003年毕业于华中科技大学建筑设计及其理论专业,获硕士学位。2003年加入北京市建筑设计研究院UFo工作室。

蔡明
UFo工作室主任建筑师,高级建筑师,国家一级注册建筑师。2003年毕业于北京建筑大学历史及其理论专业,获硕士学位。2003年加入北京市建筑设计研究院UFo工作室。

设计组长
吕娟、吴晶晶、杨坤

设计师
陈立维、陈卓群、郝一涵、李培先子、李玉洁、刘利、马思端、缪一新、石璐、孙月恒、吴锡、温琳琳、王丹妮、王风涛、王钊、尤明、赵静、周泽渥、周芸

秘书
郭瑾、孙蕾

前员工
陈莹、池胜锋、丁明达、范楷、国夫、顾知春、金国红、刘鹏飞、刘延川、闵盛勇、Nathanael Paul Maschke、潘辉、秦超、孙刚、孙晓宁、苏波、童佳旎、王承、王海亮、王舒展、王天易、王宇、卫江、肖立春、晓帆、殷霄雷、袁大伟、张士伟

Design Directors
Chen Ying
Design Director of UFo Studio, Senior Architect, National First-Class Registered Architect. She got her Master Degree in Architecture Design and Theory from the same university in 2003, and in the same year she joined BIAD UFo.

Cai Ming
Design Director of UFo Studio, Senior Architect, National First Class Registered Architect. She graduated from the Department of History and Theory of Beijing University of Civil Engineering and Architecture in 2003 with a Master's Degree. She joined BIAD UFo in 2003.

Team Leader
Lv Juan, Wu Jingjing, Yang Kun

Designers
Chen Liwei, Chen Zhuoqun, Hao Yihan, Li Peixianzi, Li Yujie, Liu Li, Ma Siduan, Miao Yixin, Shi Lu, Sun Yueheng, Wu Xi, Wen Linlin, Wang Danni, Wang Fengtao, Wang Zhao, You Ming, Zhao Jing, Zhou Zewo, Zhou Yun

Secretaries
Guo Jin, Sun Lei

Previous Members
Chen Ying, Chi Shengfeng, Ding Mingda, Fan Kai, Guo Fu, Gu Zhichun, Jin Guohong, Liu Pengfei, Liu Yanchuan, Min Shengyong, Nathanael Paul Maschke, Pan Hui, Qin Chao, Sun Gang, Sun Xiaoning, Su Bo, Tong Jiani, Wang Cheng, Wang Hailiang, Wang Shuzhan, Wang Tianyi, Wang Yu, Wei Jiang, Xiao Lichun, Xiao Fan, Yin Xiaolei, Yuan Dawei, Zhang Shiwei

图书在版编目（CIP）数据

BIAD UFo建筑工作室 / BIAD UFo工作室，群岛工作室编.
——上海：同济大学出版社，2015.1
ISBN 978-7-5608-5698-8
Ⅰ.①B… Ⅱ.①B… ②群… Ⅲ.①建筑设计－作品集－中国
－现代 Ⅳ.①TU206
中国版本图书馆CIP数据核字(2015)第277267号

BIAD UFo建筑工作室
BIAD UFo工作室，群岛工作室 编

出品人：支文军
策划：群岛工作室, Max Office, UFo工作室
项目统筹：邵韦平，李淦，刘宇光，秦蕾
项目执行：李培先子
平面设计：Max Office
（章寿品，韩建伟，王超，李骁）
责任编辑：秦蕾
特约编辑：晁艳
责任校对：徐春莲
英文校对：杨碧琼，BIAD
（陈立维，郝一涵，李玉洁，Natalie Bennett，王丹妮，周泽渥）
版 次：2015年1月第1版
印 次：2015年1月第1次印刷
印 刷：上海中华商务联合印刷有限公司
开 本：787mm×1092mm 1/16
印 张：19.5
字 数：486 000
ISBN：978-7-5608-5698-8
定 价：260.00元

出版发行：同济大学出版社
地 址：上海市杨浦区四平路1239号
邮政编码：200092
网 址：www.tongjipress.com.cn
经 销：全国各地新华书店
本书若有印刷质量问题，
请向本社发行部调换。
版权所有 侵权必究

BIAD UFo

by BIAD UFo studio, Studio Archipelago

Publisher: ZHI Wenjun
Initiator: Studio Archipelago, Max Office,
UFo studio
Director: SHAO Weiping, LI Gan,
LIU Yuguang, QIN Lei
Project executive: LI Peixianzi
Graphic design: Max Office
(ZHANG Shoupin, HAN Jianwei,
WANG Chao, LI Xiao)
Editorial team: QIN Lei (editor);
CHAO Yan (contributing editor);
XU Chunlian (proofreader)
Translation proofreader: YANG Biqiong,
BIAD (CHEN Liwei, HAO Yihan,
LI Yujie, Natalie BENNETT, WANG Danni,
ZHOU Zewo)

All rights reserved
No Part of this book may be
reproduced in any manner
whatsoever without written
permission from the publisher,
except in the context of reviews.